Synthetic Aperture Radar (SAR) Waveform Design and Signal Processing

Robert Wang · Wei Wang

Synthetic Aperture Radar (SAR) Waveform Design and Signal Processing

Springer

Robert Wang
Aerospace Information Research Institute
Chinese Academy of Sciences
Beijing, China

Wei Wang
School of Electronic, Electrical
and Communication Engineering
University of Chinese Academy of Sciences
Beijing, China

ISBN 978-981-95-7548-0 ISBN 978-981-95-7549-7 (eBook)
https://doi.org/10.1007/978-981-95-7549-7

This Springer imprint is published by the registered company Springer Nature Singapore Pte Ltd.
The registered company address is: 152 Beach Road, #21-01/04 Gateway East, Singapore 189721, Singapore

Preface

We have written this book to record our experiences in waveform design and signal processing for SAR. Most of the materials has been published previously in technical literature, but is gathered together here for the first time in a single reference work.

This book reports the latest results in the study of waveform design method for the synthetic aperture radar (SAR) and the corresponding signal processing techniques. Novel research ideas and experimental verification for the SAR waveform design have been collected on this book, including SAR system model, advanced SAR waveform design method and the corresponding signal processing method, and the application representations in the SAR system etc.

The book will be of interest to university researchers, R&D engineers and graduate students in Remote Sensing who wish to learn the core principles, methods, algorithms, and applications of SAR waveform design.

The authors apologize for any errors that may be present in the book, and would be grateful to have them brought to their attention.

Beijing, China

Robert Wang
Wei Wang

Contents

1 Fundamentals of SAR 1
1.1 Introduction 1
1.2 Basic Principle 2
1.3 System Performance 4
1.3.1 Resolution 4
1.3.2 Swath Width 5
1.3.3 Azimuth Ambiguity 6
1.3.4 Range Ambiguity 7
1.3.5 System Sensitivity 8
1.4 Summary 8
References 9

2 NLFM Waveform Design and Optimization 11
2.1 Introduction 11
2.2 NLFM Waveform Design with General Weighting Function 13
2.2.1 NLFM Waveform Based on POSP 13
2.2.2 NLFM Waveforms with Weighting Function 15
2.3 NLFM Waveform Design Based on Modified Chebyshev Weighting Function 19
2.3.1 Edge Distortion in Spectrum and Modified Chebyshev 20
2.3.2 Simulated NLFM with Modified Chebyshev 24
2.4 Real-Time Generator for NLFM Waveform 25
2.4.1 Algorithm Description 25
2.4.2 Hardware Architecture 29
2.4.3 Experimental Result 31
2.5 NLFM Waveform Optimization Method 34
2.5.1 NLFM Waveform Descriptions 34
2.5.2 Augmented Lagrangian Genetic Algorithm 36
2.5.3 Simulation Results 37
2.6 Summary 39
References 41

3 The Application of NLFM Waveform in SAR 43
3.1 Introduction 44
3.2 SNR Loss Problem 45
3.3 Matched Filtering of NLFM Waveform 46
3.4 Imaging Algorithms and Doppler Compensation for NLFM-SAR System 49
3.4.1 Modified Range Migration Algorithm 49
3.4.2 Modified Chirp Scaling Algorithm 53
3.4.3 Compensation for Doppler Effect ON NLFM SAR Imaging 58
3.5 Real SAR System and Airborne Experiments 61
3.5.1 Real SAR System for the NLFM Waveform 61
3.5.2 Airborne SAR Experiment 66
3.6 Summary 71
References 72

4 Quasi-Orthogonal Waveform Design for Range Ambiguity Suppression 75
4.1 Introduction 75
4.1.1 Background and Previous Works 75
4.1.2 Main Contribution and Organization 77
4.2 Quasi-Orthogonal Waveform Design 78
4.2.1 Signal Model 78
4.2.2 Mitigating Range Ambiguities with Quasi-Orthogonal NLFM Waveforms 80
4.3 Transmission Scheme and Signal Model 83
4.3.1 Signal Model 83
4.3.2 Imaging Performance 84
4.3.3 Ambiguity Suppression Performance 86
4.4 Signal Processing 87
4.5 Simulation Experiment 89
4.5.1 Point Target Simulation 89
4.5.2 Distributed Target Simulation 93
4.6 Summary 96
4.7 The Derivation of (4.34) 97
References 97

5 Azimuth Phase Code for Ambiguity Suppression 101
5.1 Introduction 101
5.2 Azimuth Phase Code Method 103
5.2.1 APC Modulation and Demodulation 103
5.2.2 APC Gain for Conventional SAR System 106
5.3 The Application of APC in Multichannel SAR 109
5.3.1 Signal Model with APC in Multichannel SAR 109

5.3.2 Ambiguity Suppression Based on APC in a Multichannel SAR System 113
5.4 Reconstruction Performance of the Proposed Algorithm 125
5.4.1 Reconstruction Performance Versus PRF 125
5.4.2 Distributed Scene Simulation 129
5.5 Summary 130
References 131

6 Ambiguity Suppression Scheme for Quad-Pol SAR Based on Quasi-Orthogonal Waveforms 133
6.1 Introduction 133
6.1.1 Background and Previous Works 133
6.1.2 Main Contribution and Organization 135
6.2 The Design of Quasi-Orthogonal NLFM Waveforms 136
6.2.1 Signal Model 136
6.2.2 Correlation Function 137
6.2.3 Optimization Method 141
6.3 Conventional Quad-Pol SAR 143
6.3.1 Ambiguity Performance Using Quasi-Orthogonal NLFM Waveforms 143
6.3.2 Extension Using a Combination of Quasi-Orthogonal NLFM Waveforms and APC 147
6.4 Simulation and Performance Analysis 149
6.4.1 Waveform Performance Evaluation 149
6.4.2 Point Target Simulation 149
6.4.3 Distributed Target Simulation 155
6.5 Conclusion 158
6.6 The Derivation of (6.19) 160
References 161

7 MIMO-SAR Waveform Design 163
7.1 Introduction 163
7.1.1 Previous Work 164
7.1.2 Main Advantages and Organization 165
7.2 MIMO-SAR Echo Separation Scheme 166
7.2.1 Simple Overview of the Echo Separation Problem 166
7.2.2 Scheme Description 168
7.2.3 Spatial Filter Design 174
7.3 Performance Analysis 176
7.3.1 Swath Width 177
7.3.2 SNR Performance 177
7.3.3 The Number of Band-Pass Filters 180
7.4 Simulation Experiment 180
7.5 Comparisons 184
7.5.1 SPC Waveform Versus OFDM Waveform 184

7.5.2 Comparisons Between Different MIMO-SAR Schemes 187
7.5.3 Further Discussion Based on Functional Analysis Theory 188
7.6 Summary 192
References 192

Chapter 1
Fundamentals of SAR

1.1 Introduction

Since the 1930s, radar, as an active electromagnetic sensor, has been widely used and developed for detecting, imaging, and locating reflective objects [1]. The target detection accuracy of traditional radar primarily depends on the physical aperture of the antenna, which limits the resolution performance of radar in the vertical direction of the beam. To overcome this limitation of traditional radar, Carl Wiley from Goodyear Aerospace in the United States first proposed the concept of synthetic aperture radar (SAR) in 1951 [2].

According to the type of platform, SAR can be categorized into airborne SAR and spaceborne SAR [3, 4]. One of the advantages of airborne platforms is high flexibility, which facilitates the validation of SAR principles, equipment debugging, and the development of imaging algorithms. However, airborne SAR inevitably faces certain challenges. On the one hand, airborne platforms are limited by their flight altitude and restricted by airspace sovereignty, resulting in a limited swath width of imaging for airborne SAR. On the other hand, the stability of airborne platforms is relatively poor, and airborne SAR typically observes a wide variation in look angle, which can lead to degradation of imaging quality. In contrast, SAR mounted on satellite platforms, space stations, or space shuttles can address the aforementioned problems. Spaceborne SAR operates in space, with satellite orbits typically reaching hundreds to thousands of kilometers in altitude, resulting in a considerable distance between the radar and the imaging target. The notable advantages of spaceborne SAR include high stability of platform, wide swath imaging, and freedom from sovereign airspace restrictions. These features make it indispensable in applications such as ocean observation and global terrain mapping. Consequently, since 1978, when NASA's Jet Propulsion Laboratory (JPL) launched the first ocean satellite equipped in world with a SAR sensor [5], spaceborne SAR has undergone rapid development over the past four decades. Inevitably, spaceborne SAR requires

R. Wang and W. Wang, *Synthetic Aperture Radar (SAR) Waveform Design and Signal Processing*, https://doi.org/10.1007/978-981-95-7549-7_1

higher transmission power and more advanced antenna performance to meet imaging requirements, resulting in significantly higher costs compared to airborne SAR.

SAR leverages the motion of its platform to scan the target from multiple angles and collect the corresponding echo signals [6, 7]. By synthesizing these received signals, a virtual long aperture is created, which is equivalent to a much longer physical antenna being applied. This significantly enhances the azimuth resolution of radar, freeing azimuth resolution from the limitations imposed by the physical size of the antenna. Therefore, one of the advantages of SAR is its ability to achieve high-resolution imaging of the observed area by utilizing wideband pulse signals in range and Doppler frequency signals in azimuth. In addition, as an important microwave remote sensing technology, SAR leverages the unique penetration capability of microwave frequencies to operate reliably under adverse weather conditions, at night, and through cloud cover. This enables SAR to provide global, all-time, and all-weather Earth observation capabilities. As a result, SAR is able to image targets that are difficult to detect with passive remote sensing equipment such as ordinary optical and hyperspectral sensors. With above unique advantages, SAR has been widely applied in the past few decades in fields such as terrain mapping, geological exploration, ocean observation, agricultural and forestry monitoring, and disaster assessment. In the future, SAR continues to evolve toward higher resolution, better performance, and broader applications [8].

This chapter first introduces the basic principle of SAR, followed by a review and explanation of some key performance aspects of SAR systems. Finally, the content and structure of this chapter are briefly summarized.

1.2 Basic Principle

During radar mapping, the direction parallel to the flight path is typically referred to as the azimuth, while the direction perpendicular to the flight path is referred to as the range. Assume that the radar wavelength is λ, and the slant range between the target and the radar platform is R. For an antenna with a real aperture length of L_a, the beamwidth of the main lobe is approximately λ/L_a.

In conventional radar, the azimuth resolution can be calculated as:

$$\rho_a = R\frac{\lambda}{L_a} \tag{1.1}$$

It is evident that the longer the antenna, the narrower the radar beam and the higher the azimuth resolution. However, in practical applications, it is not feasible to have an extremely large antenna size, which limits the improvement of azimuth resolution in conventional radar systems. In addition, the azimuth resolution of radar is directly proportional to the slant range between the target and the platform. Therefore, the azimuth resolution deteriorates as the slant range between the target and the platform increases.

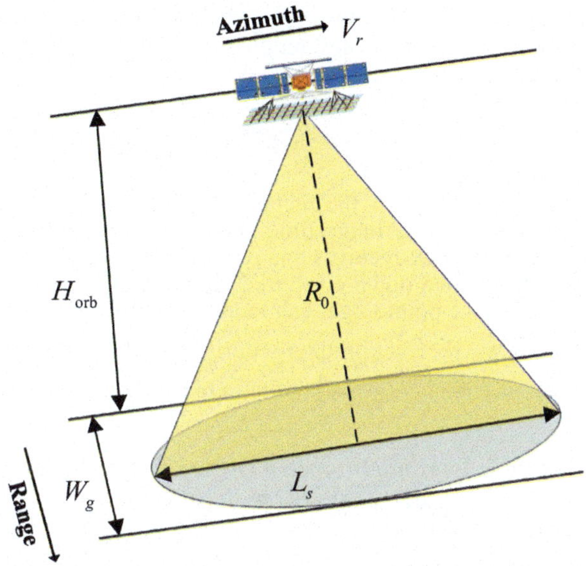

Fig. 1.1 Schematic of SAR geometric model

SAR has been proposed to mitigate the above problems, which operates in various modes, such as stripmap SAR mode, scanSAR mode, and spotlight SAR mode [7, 9, 10]. Taking the traditional single-channel stripmap SAR mode as an example, this mode refers to a scenario where the SAR beam remains fixed in direction as the platform moves, illuminating a strip-shaped area on the ground. The geometric relationship during operation is illustrated as. In Fig. 1.1, the SAR mounted on the platform moves parallel to the ground at a constant velocity V_r. For simplicity, the orthogonal side-looking geometry model is considered, where the SAR beam is strictly perpendicular to the motion direction of SAR. The pulse signals are usually transmitted by SAR at a constant pulse repetition frequency (PRF), and then the coherent signals are received and processed. This is equivalent to coherently summing the signals transmitted by antenna elements arranged with a spacing of $\Delta x = V_r/\text{PRF}$. As the SAR platform moves forward, an equivalent synthesized linear antenna array is formed, with the synthetic aperture length given by:

$$L_s = R\frac{\lambda}{L_a} \tag{1.2}$$

The synthetic aperture length represents the trajectory length traversed by the SAR beam while illuminating a target, and the corresponding flight time is referred to as the synthetic aperture time. Since the synthetic aperture length L_s is much larger than the real antenna aperture L_a, the synthetic beamwidth is significantly smaller than the real beamwidth, thereby greatly enhancing the azimuth resolution of SAR.

During the mapping process of SAR, the platform moves along the azimuth, and the SAR beam transmits phase-modulated pulse signals in a direction nearly perpendicular to the motion of the platform [7]. Assuming that the SAR transmits an ideal linear frequency-modulated signal, which is expressed as:

$$s(t) = \mathrm{rect}\left(\frac{t}{T_r}\right)\exp\left\{j\pi K_r t^2\right\} \tag{1.3}$$

where rect (t/T_r) is the rectangular function of width T_r in range. T_r is the pulse duration, t is the range time and K_r is the range frequency-modulated rate. The receiver of SAR receives and records the echoes reflected from the Earth's surface, which are then down-converted, demodulated, and digitized through sampling and storage. The point target echo model can be expressed as:

$$s(t,\eta) = \mathrm{rect}\left(\frac{t-\frac{2R(\eta)}{c}}{T_r}\right)\cdot w_a(\eta)\cdot\exp\left\{-j4\pi f_c\frac{R(\eta)}{c}\right\}\cdot\exp\left\{j\pi K_r\left(t-\frac{2R(\eta)}{c}\right)^2\right\} \tag{1.4}$$

where η is azimuth time, $R(\eta)$ is instantaneous slant range and can be written as $R(\eta) = \sqrt{R_0{}^2 + (V_r\eta)^2}$, R_0 is minimum slant range. $w_a(\eta)$ is azimuth antenna pattern, f and c are the carrier frequency and velocity of light respectively. It can be seen that the echo delay is directly proportional to the slant range between the radar and the scatterer. In range, by performing pulse compression on the echo signal, output pulses much narrower than the transmitted pulses can be obtained, resulting in higher range resolution. In azimuth, the radar platform continuously transmits signals during its flight, and after coherent processing of the received echoes, it is equivalent to a synthesized array antenna with a significantly increased length, thereby improving the azimuth resolution [7, 11]. Ultimately, a two-dimensional grayscale SAR image of the observed area can be obtained.

1.3 System Performance

SAR is not an ideal sensor, and its imaging performance is limited by system characteristics and inherent constraints. This section primarily introduces several key metrics for evaluating the performance of SAR systems from the perspective of traditional single-channel systems, including: geometric resolution, swath width, azimuth ambiguity, range ambiguity, and system sensitivity [6, 12].

1.3.1 Resolution

According to SAR imaging theory, the imaging result of an ideal point target can be approximated as a two-dimensional sinc function. The geometric resolution of a SAR image is defined as the half-power width of the main lobe in the point target imaging result. The resolution represents the ability of SAR to distinguish between two adjacent targets on the ground [7].

Based on the principle of SAR imaging, considering the shared use of the SAR antenna for both transmission and reception, and the two-way transmission of electromagnetic waves between the antenna and the target, which induces a two-way phase shift in azimuth, the azimuth resolution of SAR can be given by:

$$\rho_a = \frac{1}{2} R \frac{\lambda}{L_s} = \frac{L_a}{2} \tag{1.5}$$

It is evident that, theoretically, the azimuth resolution of SAR is only related to the antenna length and is independent of factors such as wavelength and slant range. The shorter the antenna length, the better the azimuth resolution.

The range resolution is determined by the bandwidth of the SAR transmitted signal and is classified into slant range resolution and ground range resolution. Based on the SAR imaging geometry, the slant range resolution and ground range resolution can be expressed as

$$\rho_r = \frac{c}{2B_r} \tag{1.6}$$

and

$$\rho_g = \frac{c}{2B_r \sin(\alpha)} \tag{1.7}$$

where α is the incidence angle of SAR beam and B_r is the transmitted bandwidth of signal. It can be seen that for a SAR system, the slant range resolution is a constant and is solely determined by the transmitted bandwidth, while the ground range resolution depends on the incidence angle and varies across the entire mapping swath.

1.3.2 Swath Width

Swath width is one of the most important performance parameters of a SAR system and defined as the width of the ground range corresponding to the area illuminated by the beam during SAR operation [7].

Assuming that the incidence angle at the center of the interested swath is α_c, corresponding to a slant range of R_c, the height of the antenna is L_h and the beamwidth of SAR in range is θ_e. Then, the swath widths in ground range and slant range can be expressed as

$$W_g = \frac{\theta_e \cdot R_c}{\cos(\alpha_c)} = \frac{\lambda \cdot R_c}{L_h \cdot \cos(\alpha_c)} \tag{1.8}$$

and

$$W_r = \frac{\lambda \cdot R_c \cdot \sin(\alpha_c)}{L_h \cdot \cos(\alpha_c)} \tag{1.9}$$

It can be seen that the wider the beamwidth in range, the wider the corresponding swath. When the beamwidth is fixed, the farther the illuminated area, the larger the incidence angle, and the wider the mappable swath.

1.3.3 Azimuth Ambiguity

The slant range trajectory of the SAR azimuth beam main lobe when illuminating the target affects the phase of the echo signal, forming the azimuth Doppler frequency signal. When other parameters of the SAR system are fixed, the Doppler bandwidth is determined by the beamwidth of the azimuth antenna pattern. The SAR beamwidth is typically defined as the 3 dB main lobe width of the antenna pattern. However, due to the non-ideal nature of the antenna beam, some energy outside the beamwidth still reaches the target, which results in the actual Doppler bandwidth being larger than the Doppler bandwidth determined by the 3 dB main lobe width. From the sampling perspective, the echo of SAR is sampled at the PRF in azimuth and the azimuth signal is discrete. This requires the azimuth signal to be band-limited, with a bandwidth smaller than the PRF. However, the Doppler bandwidth of the azimuth signal is not strictly limited to B_d in practice. Therefore, when sampling at the PRF, signals with Doppler frequencies that are integer multiples of the PRF in the echo will be extended into the main lobe, resulting in aliasing of the main lobe signal, which causes azimuth ambiguity in SAR [13]. Figure 1.2 gives a schematic of azimuth ambiguity, where f_{dc} is the Doppler frequency center, f_η is the Doppler frequency, and w_a is the azimuth antenna pattern.

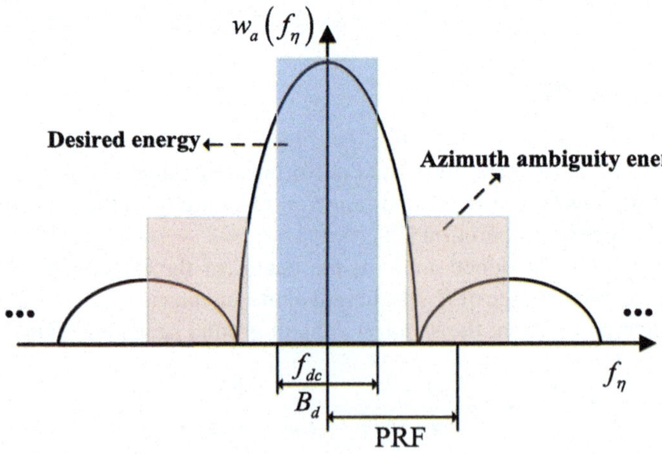

Fig. 1.2 Schematic of azimuth ambiguity, where the horizontal axis represents the Doppler frequency and the vertical axis represents the azimuth antenna pattern

Azimuth ambiguity can lead to false targets during SAR imaging, which negatively impacts the final image quality. This performance is typically assessed using the azimuth ambiguity-to-signal ratio (AASR). Based on the above principle, the equation for estimating the AASR of a SAR system can be expressed as

$$AASR = \frac{\sum\limits_{p=-\infty,a\neq 0}^{p=+\infty} \int_{f_{dc}-B_d/2}^{f_{dc}+B_d/2} w_a{}^2\left(f_\eta + p \cdot \mathrm{PRF}\right) \mathrm{d} f_\eta}{\int_{f_{dc}-B_d/2}^{f_{dc}+B_d/2} w_a{}^2\left(f_\eta\right) \mathrm{d} f_\eta} \tag{1.10}$$

where p is the order of azimuth ambiguity. In above formula, the numerator represents the ambiguity energy, and the denominator represents the signal energy. Therefore, the smaller the AASR, the less the radar image is affected by ambiguity energy, indicating better azimuth ambiguity performance of the system.

1.3.4 Range Ambiguity

The range antenna pattern of SAR is also not ideal, and the actual coverage width of the SAR beam in range is broader than the swath of interest for mapping, leading to echo energy being returned from outside the mapped swath that is illuminated by the main lobe. The echoes from outside the swath of interest mix with the expected echoes from the swath of interest. After signal processing, this results in a degradation of SAR image quality, affecting the identification of real targets. The energy from outside the mapped swath is referred to as range ambiguity, and the formation of range ambiguity is schematically illustrated in Fig. 1.3.

The range ambiguity-to-signal ratio (RASR) is typically used to evaluate the range ambiguity performance of a SAR system. It is defined as the ratio of the echo energy from the ambiguity region to the echo energy from the interested swath. The energy expression can be derived from the radar equation, and the corresponding RASR evaluation formula is as follows

$$RASR = \frac{\sum\limits_{q=-\infty,q\neq 0}^{q=+\infty} \frac{\sigma\left(\alpha_{amb,q}\right)\cdot G_t\left(\varphi_{amb,q}\right)\cdot G_r\left(\varphi_{amb,q}\right)}{R_{amb,q}\cdot \sin\left(\alpha_{amb,q}\right)}}{\frac{\sigma\left(\alpha_0\right)\cdot G_t\left(\varphi_0\right)\cdot G_r\left(\varphi_0\right)}{R_0\cdot \sin(\alpha_0)}} \tag{1.11}$$

where q is the order of range ambiguity and the corner markers 0 and amb denote the parameters corresponding to the desired and range ambiguity signals, respectively. α and φ are the incidence angle and look angle, respectively. G_t and G_r are the transmission gain and reception gain of SAR respectively. In particular, $R_{\mathrm{amb}} = R_0 + q/\mathrm{PRF}$. It can be seen that the smaller the RASR, the less the SAR image is affected by range ambiguity energy.

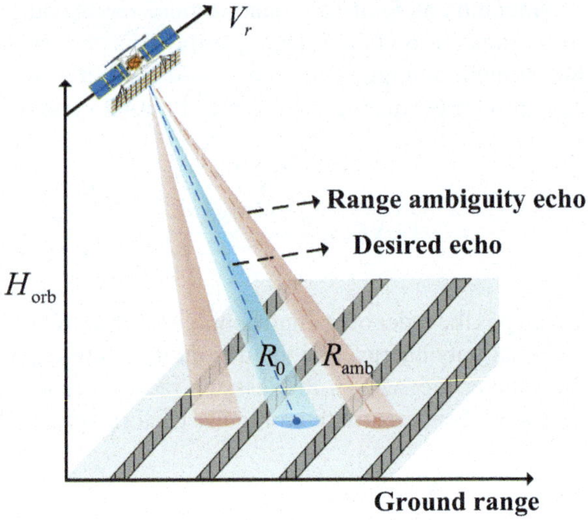

Fig. 1.3 Schematic of range ambiguity, where the horizontal axis represents the ground range, the vertical axis represents the orbit height, the upper axis of the figure shows the direction of platform movement, and the black stripes are the blind ranges caused by the transmission blockages

1.3.5 System Sensitivity

The sensitivity of a SAR system, also known as the noise equivalent sigma zero (NESZ), is used to measure the ability of SAR to image weak targets. It represents the corresponding backscatter coefficient of the target when the output signal-to-noise ratio (SNR) of SAR image is 0 dB [14]. The evaluation expression for NESZ is as follows

$$NESZ = \frac{256\pi^3 R_0 \cdot V_r \cdot \sin(\alpha_0) \cdot k_b \cdot T_n \cdot B_n \cdot L_f}{P_t \cdot \mathrm{PRF} \cdot \tau_p \cdot G_t(\varphi_0) \cdot G_r(\varphi_0) \cdot \lambda^3 \cdot c} \tag{1.12}$$

where k_b is Boltzmann's constant, T_n is the system noise temperature, B_n is the equivalent noise bandwidth of SAR system and $B_n \approx B_r$. L_f is the system loss, and P_t is the peak power. The smaller the NESZ value, the lower the threshold of the backscatter coefficient for detectable targets, indicating stronger imaging capability for weak targets and a higher signal-to-noise ratio (SNR) in the final radar image.

1.4 Summary

This chapter primarily reviews the fundamental geometric imaging principles and signal models of SAR. It begins by explaining the key concepts of SAR imaging, including the geometric principles of signal acquisition and the processing techniques required to generate high-resolution SAR images. Subsequently, the chapter introduces several important system parameters that describe the performance of SAR systems, such as SAR resolution, sensitivity, and ambiguity ratios. These parameters are crucial for understanding the capabilities and limitations of SAR systems in various environments. The discussion provides a comprehensive foundation for the subsequent chapters, laying the groundwork for a deeper exploration of advanced SAR system design, performance optimization, and application-specific considerations.

References

1. Skolnik MI. Introduction to radar. Radar handbook. 1962;2:21.
2. Wiley CA. Synthetic aperture radars. IEEE Transactions on Aerospace and Electronic Systems. 1985;(3):440–443.
3. Horn R, Nottensteiner A, Scheiber R. F-SAR-DLR's advanced airborne SAR system onboard DO228. In: 7th European Conference on Synthetic Aperture Radar. VDE; 2008. p. 1–4.
4. Werninghaus R, Buckreuss S. The TerraSAR-X mission and system design. IEEE Transactions on Geoscience and Remote Sensing. 2009;48(2):606–614.
5. Elachi C. Radar images of the earth from space. Scientific American. 1982;247(6):54–61.
6. McDonough RN. Synthetic aperture radar: systems and signal processing. Wiley; 1991.
7. Cumming IG, Wong FH. Digital processing of synthetic aperture radar data. Artech house. 2005;1(3):108–110.
8. Yunkai D, Weidong Y, Heng Z, et al. Forthcoming spaceborne SAR development. Journal of Radars. 2020;9(1):1–33.
9. Martinez A, Marchand JL. SAR image quality assessment. Revista de Teledeteccion. 1993;2:12–18.
10. Jakowatz CV, Wahl DE, Eichel PH, et al. Spotlight-mode synthetic aperture radar: a signal processing approach: a signal processing approach. Springer Science & Business Media; 2012.
11. Soumekh M. Synthetic aperture radar signal processing. vol. 7. New York: Wiley; 1999.
12. Song R, Wang W, Yu W. The Latest Developments in Spaceborne High-Resolution Wide-Swath SAR Systems and Imaging Methods. Sensors (Basel, Switzerland). 2024;24(18):5978.
13. Li F, Johnson W. Ambiguities in spacebornene synthetic aperture radar systems. IEEE Transactions on Aerospace and Electronic Systems. 1983;(3):389–397.
14. Massonnet D, Souyris JC. Imaging with synthetic aperture radar. EPFL press; 2008.

Chapter 2
NLFM Waveform Design and Optimization

Generally, radar system employ linear frequency modulation (LFM) signal to achieve the high-resolution image and weighted windowing is usually employed to suppress sidelobes. However, it usually leads to an additional 1–2 dB loss in signal-to-noise ratio (SNR). Nonlinear frequency modulation (NLFM) signal, whose power spectral density (PSD) can be shaped to reduce sidelobes without incurring SNR loss, therefore constitute a promising alternative, but it often broadens the main lobe. Also, the real-time generation of precise NLFM signal is a technical challenge. In this chapter, firstly, a novel NLFM waveform with narrower mainlobe based on the modified Chebyshev window is presented. Subsequently, we develop a high-precision NLFM signal generator with predistortion compensation capability, which has been deployed in the LT-1 (LuTan-1, i.e. TwinSAR-L) mission, an innovative spaceborne bistatic SAR system. Finally, we present a new waveform optimization framework, by which an advanced NLFM waveform with lower sidelobes and a more compact mainlobe is obtained.

2.1 Introduction

Linear frequency modulation (LFM) signals, which offer both high range resolution and favorable Doppler tolerance, are extensively deployed in radar systems. However, their correlation function inherently exhibits relatively high range sidelobes at about −13.26 dB, a level typically regarded as excessive because it can obscure weaker neighboring targets in the reconstructed image. Therefore, additional sidelobe-suppression techniques must be employed to mitigate this effect. Generally, weighted windowing [1] is employed to suppress sidelobes. However, the introduction of an additional weighting filter causes a mismatch between the filter response and the transmitted signal, leading to an SNR loss at the MF output.

Compared with the conventional LFM waveform, a nonlinear frequency modulation (NLFM) waveform can tailor the time–frequency law to synthesize a desired

R. Wang and W. Wang, *Synthetic Aperture Radar (SAR) Waveform Design and Signal Processing*, https://doi.org/10.1007/978-981-95-7549-7_2

PSD, thus yielding a MF output with reduced sidelobes while maintaining the SNR. Considerable research has been devoted to the design and generation of NLFM waveforms [2–4]. Many existing synthesis approaches are built on the principle of stationary phase (POSP) [5–8], which links the PSD to the instantaneous chirp rate at each frequency and thereby shapes the spectral distribution of the waveform. In [9], we propose a new NLFM waveform with reduced sidelobes based on a modified Chebyshev window.

Although NLFM waveforms can achieve sidelobe reduction without SNR loss, their real-time generation is technically more challenging than that of conventional LFM signals. The Institute of Electronics, Chinese Academy of Sciences (IECAS) carried out the first airborne SAR experiment using an NLFM signal [10]; however, in that system the transmitted NLFM waveform was pre-stored in the FPGA ROM. In practical applications, the signal generator must be capable of producing waveforms with varying parameter sets, and it is evident that a data-storage-based scheme is infeasible due to the limited ROM capacity. Consequently, there is an urgent demand for simple and accurate waveform generation algorithms suitable for real-time computation. Several concise analytic signal models have been proposed, such as the model formed by combining an LFM signal with a tan function [11], an empirical NLFM model (Eq. (5.20)) [12], as well as the time Exponentiated Frequency Modulation (TEFM) and Sine Exponentiated Frequency Modulation (SEFM) models [13]. However, compared with weighted LFM signals, the matched-filter output performance of these models, including PSLR and mainlobe width (MW), is typically degraded. A more favorable approach is to shape the spectrum with a window designed via POSP, but such methods are usually too complicated and computationally intensive for real-time implementation. In this chapter, a modified NLFM generation method based on POSP is proposed [14], which has been integrated as a unit of the signal source for the LT-1 (LuTan-1, i.e. TwinSAR-L) mission.

Building on the above approaches, NLFM waveforms can suppress sidelobes without incurring SNR loss, but this advantage is usually accompanied by a broadening of the main lobe. Hence, the pursuit of an optimal NLFM waveform design remains an ongoing topic of research. An NLFM optimization method based on the B’ezier curve was reported in [15]. In [16], Saeedi et al. employed generalized piecewise linear (PWL) functions to parameterize a symmetric instantaneous frequency law and adopted a multi-objective goal-attainment optimization algorithm to optimize the waveform. More recently, continuous phase modulation (CPM) waveforms have attracted considerable interest. Within the CPM framework, Jakabosky et al. [17] designed waveforms using Marginal Fisher’s Information, achieving sidelobe reduction without mainlobe broadening. They subsequently proposed a spectral-shaping strategy that jointly optimizes the FM component of the waveform and a low-loss amplitude taper via spectral shaping optimization [18]. Experimental results showed range sidelobes below −80 dB with less than 0.3 dB SNR loss. Blunt et al. [19] further utilized a greedy search procedure, exploiting complementary metrics such as peak sidelobe level and integrated sidelobe level to refine the waveform design. However, whether the signal is an FM or a CPM waveform, a parametric description—such as a B’ezier curve, PWL functions, or phase codes—is first required, and an optimization algorithm is then applied to search for a suitable waveform. In this chapter, we present a new optimization framework for NLFM waveforms based on the Augmented Lagrangian Genetic Algorithm (ALGA) and PWL functions [20]. Asymmetric PWL functions are used to define the waveform search space, while ALGA is employed to carry out the waveform optimization.

This chapter is organized as follows. Section 2.2 reviews the POSP-based synthesis of NLFM waveforms and introduces several NLFM designs associated with different weighting functions. Section 2.3 then presents a modified Chebyshev weighting function for generating a novel NLFM waveform with reduced sidelobes [9]. In Sect. 2.4, an advanced NLFM generation scheme that markedly decreases the computational burden for real-time implementation is described. Section 2.5 details the optimization of NLFM waveforms using generalized piecewise linear (PWL) functions together with the Augmented Lagrangian Genetic Algorithm (ALGA). Finally, Sect. 2.6 provides concluding remarks for this chapter.

2.2 NLFM Waveform Design with General Weighting Function

Since the ACF is the inverse Fourier transform of the power spectrum, its sidelobes can be suppressed by appropriately shaping the spectrum using sidelobe-reduction weighting windows. The method for shaping the power spectrum is to generate NLFM waveform. According to the POSP, for an NLFM signal with a large time bandwidth product, the power spectral density at a given frequency is inversely proportional to the chirp rate at that frequency [12, 20]. In this section, we review the POSP-based procedure for generating NLFM waveforms, which has a specific chirp rate at a particular frequency point to attain the predefined power spectrum.

2.2.1 *NLFM Waveform Based on POSP*

For a radar system, an NLFM baseband waveform with bandwidth B_r can in general be written as

$$s(t) = \mathrm{rect}\left(\frac{t}{T_r}\right)\exp\left(\mathrm{j}\theta(t)\right) \tag{2.1}$$

with boundary constraints

$$\theta'(T/2) = \pi B_r \quad \theta'(-T/2) = -\pi B_r \tag{2.2}$$

where t denotes the range time, T_r is the pulse duration, $\theta(t)$ is the signal phase, and $\mathrm{rect}(\cdot)$ represents the rectangular window function defined as

$$\mathrm{rect}(x) = \begin{cases} 1, & |x| <= 0.5 \\ 0, & |x| > 0.5 \end{cases} \tag{2.3}$$

Further, the spectrum $S(f)$ of the NLFM waveform can be written as

$$S(f) = \int_{-\infty}^{\infty} \mathrm{rect}\left(\frac{t}{T_r}\right) e^{\mathrm{j}\theta(t) - \mathrm{j}2\pi f t} dt \tag{2.4}$$

where f is the frequency index. According to [21], the spectrum $S(f)$ can be approximated as

$$S(f) \approx \mathrm{rect}\left(\frac{f}{B_r}\right) e^{\mathrm{j}\theta(t) - \mathrm{j}2\pi f t} \sqrt{\frac{2\pi}{\theta''(t)}} \tag{2.5}$$

and $\theta(t)$ satisfies

$$\frac{d\theta(t)}{dt} = 2\pi f \tag{2.6}$$

Indeed, the power spectrum is written by

$$P(f) = |S(f)|^2 \approx \mathrm{rect}\left(\frac{f}{B_r}\right) \frac{2\pi}{\theta''(t)} = \frac{dt}{df} \tag{2.7}$$

According to (2.7), the power spectrum of the NLFM waveform is therefore proportional to the derivative of the group delay function $t = t(f) = G(f)$, or inversely proportional to the derivative of the instantaneous frequency function $f(t)$. To shape the power spectrum $P(f)$ by a prescribed weighting function $W(f)$, we impose

$$G(f) = \int_{-\infty}^{f} W(f) df \tag{2.8}$$

and the corresponding instantaneous frequency law is then obtained as

$$f = f(t) = G^{-1}(t) \tag{2.9}$$

By (2.6), the phase function $\theta(t)$ of the NLFM waveform is integration of $2\pi f(t)$. Therefore, the NLFM waveform is accordingly expressed as

$$s(t) = \mathrm{rect}\left(\frac{t}{T}\right) \exp\left(\mathrm{j}2\pi \int_0^t f(t) dt\right) \tag{2.10}$$

Moreover, the power spectrum shape of this NLFM waveform closely approaches the predefined weighting function.

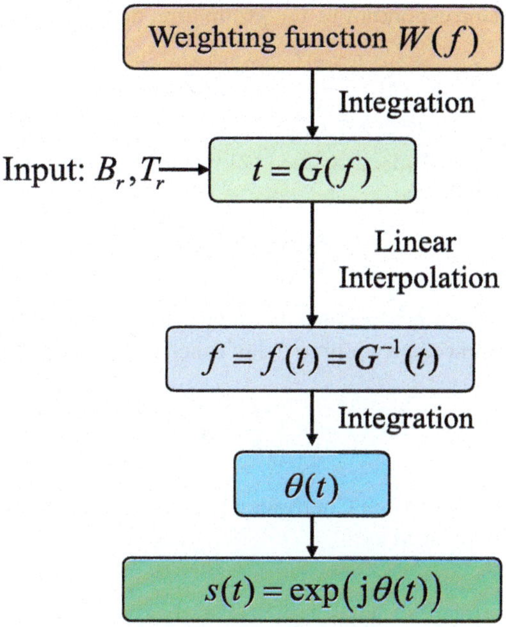

Fig. 2.1 The process of generating an NLFM waveform with predefined power spectrum shape based on the POSP

Building on the above POSP formulation, the procedure for synthesizing an NLFM waveform is summarized in Fig. 2.1. As depicted in Fig. 2.1, linear interpolation is employed to numerically calculate the instantaneous frequency from the group time-delay function. Moreover, to guarantee sufficient accuracy of the instantaneous-frequency estimate, the number of samples for the group time-delay function should be chosen considerably larger than that for the instantaneous-frequency sequence.

2.2.2 NLFM Waveforms with Weighting Function

By appropriately designing the NLFM waveform, its PSD is shaped to follow the desired weighting function $W(f)$. Since the autocorrelation function (ACF) and the PSD form a Fourier transform pair, the choice of weighting function directly governs the performance characteristics of the NLFM waveform. In the following, we briefly introduce the uniform, raised-cosine, Dolph–Chebyshev, Kaiser, Taylor, Gauss, and Hamming weighting functions together with their associated group delay functions.

(1) Uniform weighting function. The N-point coefficients of uniform weighting are expressed as

$$w(n) = 1, \ |n| \le \frac{N-1}{2} \tag{2.11}$$

Correspondingly, the group time delay function can be derived as

$$G(f) = \frac{T}{B} f, \quad |f| \leq B/2 \tag{2.12}$$

(2) Raised cosine [12] weighting function.

$$w(n) = k + (1-k)\cos^2\left(\frac{\pi n}{N-1}\right), \quad |n| \leq \frac{N-1}{2} \tag{2.13}$$

where k is a constant value which can control sidelobe level and mainlobe width, and the group time delay function is written by

$$G(f) = T\left(\frac{f}{B} + \left(\frac{1-k}{1+k}\right)\frac{1}{2\pi}\sin\left(\frac{2\pi f}{B}\right)\right), \quad |f| \leq B/2 \tag{2.14}$$

(3) Dolph Chebyshev [22] weighting function. The Dolph weighting function is expressed as

$$w_n = 1 + 2r\sum_{m=1}^{M} T_{2M}\big(x_0\cos(\theta_m/2)\big)\cos(m\theta_n) \tag{2.15}$$

where $|n| \leq M$, $N = 2M+1$, $\theta_m = 2\pi m/N$, r represents the ripple ratio, x_0 is expressed as

$$x_0 = \cosh\left(\frac{1}{2M}\cosh^{-1}\left(\frac{1}{r}\right)\right) \tag{2.16}$$

and $T_{2M}(x)$ denotes the Chebyshev polynomials, given by

$$T_n(x) = \begin{cases} \cos(n\cos^{-1}x), & |x| \leq 1 \\ \cosh(n\cosh^{-1}x), & |x| > 1 \end{cases} \tag{2.17}$$

However, the Dolph weighting function has no closed-form expression of group time delay function.

(4) Kaiser [23] weighting function.

$$w(n) = \begin{cases} \dfrac{I_0\left(\pi\alpha\sqrt{1-\left(\frac{2n}{N}\right)^2}\right)}{I_0(\pi\alpha)}, & |n| <= \dfrac{N-1}{2} \\ 0, \text{ others} \end{cases} \tag{2.18}$$

where I_0 and α denote the zero-order modified Bessel function of the first kind and an positive number, respectively. Also, the Kaiser weighting function has no closed-form expression of group time delay function.

(5) Taylor [24] weighting function.

$$w(n) = 1 + \sum_{m=1}^{\overline{n}-1} F_m \cos\left(\frac{2\pi mn}{N-1}\right), \quad |n| \leq \frac{N-1}{2} \tag{2.19}$$

where $F_m = F(m, \overline{n}, \eta)$ denotes the mth-order Taylor coefficient. Here, η specifies the sidelobe level, and $\overline{n}$ represents the number of sidelobes with equal amplitude. The group time delay function for Taylor weighting function is derived as

$$G(f) = T\left(\frac{f}{B} + \frac{1}{2\pi}\sum_{m=1}^{\overline{n}-1} \frac{F_m}{m} \sin\left(\frac{2\pi mf}{B}\right)\right), \quad |f| \leq B/2 \tag{2.20}$$

(6) Gauss [25] weighting function

$$w(n) = e^{-\frac{1}{2}\left(\frac{2\alpha n}{N-1}\right)^2}, |n| \leq \frac{N-1}{2} \tag{2.21}$$

where α controls the shape of Gauss window. There is no closed-form expression of group time delay function for Gauss weighting function.

(7) Hamming [26] weighting function

$$w(n) = 0.54 - 0.46\cos\left(\frac{2\pi n}{N-1}\right), 0 \leq n \leq N-1 \tag{2.22}$$

The group time delay function for Hamming weighting function is derived as

$$G(f) = T\left(\frac{f}{B} + \frac{0.46}{2\pi \cdot 0.54} \sin\left(\frac{2\pi f}{B}\right)\right), \quad |f| \leq B/2 \tag{2.23}$$

Figure 2.2 illustrates the coefficients of the above weighting functions in case of $N = 200$. It should be noted that only the uniform, raised cosine, Taylor and Hamming weighting function have the closed-form group time delay function, while the group time delay function of other weighting functions must be carried out numerically.

Furthermore, it can be observed from the expressions of the above weighting functions that only the instantaneous frequency of uniform weighting function can be analytically derived as

$$f(t) = \frac{B_r}{T_r} t \tag{2.24}$$

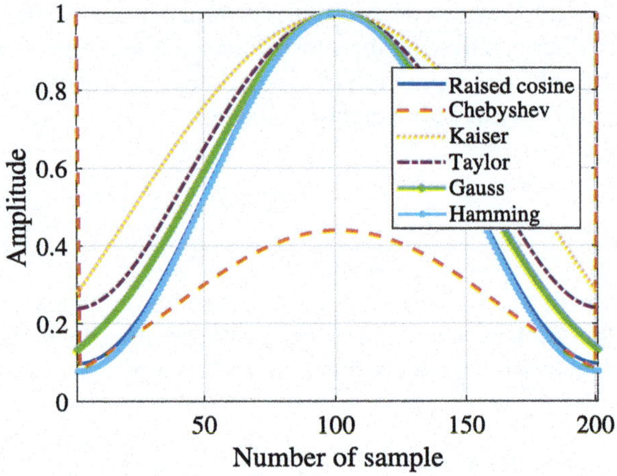

Fig. 2.2 Illustration of raised cosine, Dolph, Kaiser, Taylor, Gauss and Hamming window functions with $N = 200$

while, in general, all other weighting functions do not have closed-form expressions because these instantaneous frequency functions are very complex. Thus, these instantaneous frequency functions of the group time delay function must be carried out numerically. Figure 2.3 illustrates, as an example, the instantaneous frequency functions of NLFM waveforms (signal bandwidth: $B = 300$ MHz, pulse duration: $T = 20$ μs) that arise from the above weighting functions by using integration and linear interpolation. The ACF (close-up) of these NLFM waveforms is illustrated in Fig. 2.4. Also, the following measures are often used to evaluate the ACF performance of NLFM waveforms.

(1) Impulse Response Width (IRW): defined as the 3-dB width of the mainlobe of the ACF, i.e., the mainlobe width measured at the point where its magnitude drops by 3 dB from the peak value.

(2) PSLR: the peak sidelobe level ratio, defined as the ratio (in decibels) between the amplitude of the highest sidelobe and that of the mainlobe peak in the ACF.

(3) ISLR: it can be obtained by integrating the power of the ACF over suitable regions. If the "main lobe" power is P_{main} and the total power is P_{total}, the ISLR is then

$$\text{ISLR} = 10\log_{10}\left(\frac{P_{\text{total}} - P_{\text{main}}}{P_{\text{main}}}\right) \tag{2.25}$$

Further, the performances including IRW, PSLR, ISLR of different NLFM waveforms are summarized in Table 2.1.

From the above analysis, NLFM waveforms can tailor the time–frequency law to shape the power spectrum according to a prescribed window, thereby achieving a low-sidelobe ACF without incurring SNR loss. Consequently, the NLFM signal constitutes a viable substitute for the conventional LFM signal.

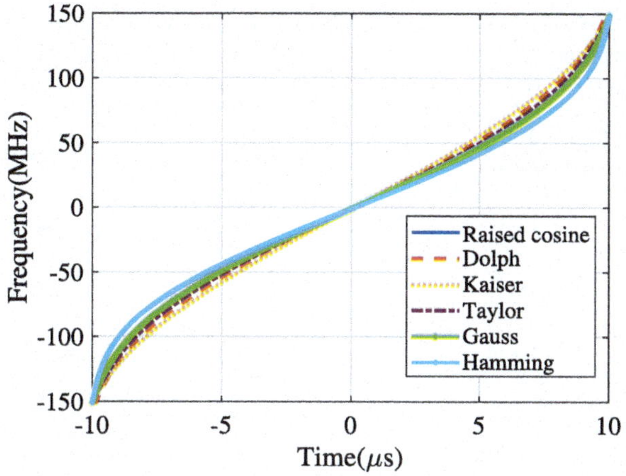

Fig. 2.3 The instantaneous frequency functions that arise from the group time delay functions of weighting functions (raised cosine, Dolph, Kaiser, Taylor, Gauss and Hamming) by using linear interpolation

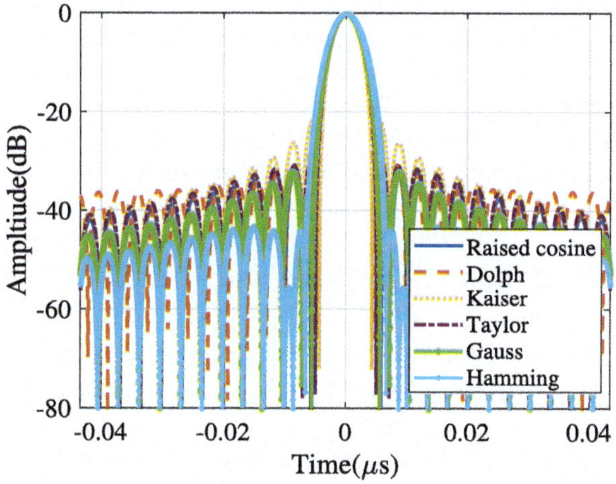

Fig. 2.4 The ACF (close-up) of the NLFM waveforms whose power spectrum are shaped by different weighting functions

2.3 NLFM Waveform Design Based on Modified Chebyshev Weighting Function

Dolph [27] showed that, for a given sidelobe level, the Chebyshev window yields the minimum impulse response width (IRW). This suggests that an NLFM waveform whose PSD follows a Chebyshev window should exhibit lower range sidelobes than those obtained with other windowed PSDs, without broadening the main lobe.

Table 2.1 ACF performances for the NLFM waveforms whose power spectrum are shaped by different weighting functions

NLFM waveform	IRW (Sample)	PSLR (dB)	ISLR (dB)
Uniform-shape (LFM)	0.886	−13.26	−9.72
Raised cosine-shape ($k = 0.2$)	1.169	−31.68	−29.21
Dolph-shape ($sll = -30$)	1.084	−27.82	−20.33
Kaiser-shape ($\alpha = 2.6$)	1.051	−21.48	−19.07
Taylor-shape ($sll = -30$)	1.124	−30.30	−23.54
Gauss-shape ($\alpha = 2$)	1.184	−31.93	−27.22
Hamming-shape	1.303	−43.06	−34.57

In practice, however, the PSLR achieved by raised-cosine and Taylor weightings is observed to be superior to that of the Chebyshev case. Two factors may account for this degradation in Chebyshev performance. First, the Chebyshev window is non-differentiable at its edges. Second, spectral approximation introduces distortions in the effective Chebyshev response. In this section, we rederive the optimized Chebyshev window and the corresponding NLFM spectrum, and then propose a modified Chebyshev (MCheb) window for NLFM waveform design [9].

2.3.1 *Edge Distortion in Spectrum and Modified Chebyshev*

In this section, the edge distortion of NLFM spectrum is firstly described. Then a modified Chebyshev weighting function is deduced to generate an NLFM waveform with low sidelobes.

In detail, according to (2.4), assuming that the derivative of the function $\varphi(t) = \theta(t) - 2\pi f t$ vanishes at only a single point $t = t_0$ in the interval $(-\infty, \infty)$:

$$\theta'(t_0) = 2\pi f \quad \theta''(t_0) \neq 0 \tag{2.26}$$

Thus, by quadratic approximation, the phase can be reformulated as

$$\varphi(t) \approx \theta(t_0) - 2\pi f t_0 + \frac{\theta''(t_0)}{2}(t - t_0)^2 \tag{2.27}$$

It follows that

$$S(f) \approx e^{(\mathrm{j}(\theta(t_0) - 2\pi f t_0))} \int_{-\infty}^{\infty} \mathrm{rect}\left(\frac{t}{T_r}\right) e^{\left(\mathrm{j}\frac{\theta''(t_0)}{2}(t-t_0)^2\right)} dt \tag{2.28}$$

Define the Fresnel integral

$$K(x) = \int_0^x e^{j\pi\tau^2/2} d\tau \tag{2.29}$$

Then

$$S(f) \approx e^{(j(\theta(t_0)-2\pi f t_0))} \sqrt{\frac{\pi}{\theta''(t_0)}} \times G(f) \tag{2.30}$$

where

$$\begin{aligned} G(f) = K\left(\sqrt{\frac{\theta''(t_0)}{\pi}}\left(\frac{T}{2} - t_0\right)\right) \\ + K\left(\sqrt{\frac{\theta''(t_0)}{\pi}}\left(\frac{T}{2} + t_0\right)\right) \end{aligned} \tag{2.31}$$

In general, the Fresnel integral term in (2.31) can be approximated as $\sqrt{2}\text{rect}(t_0/T)$, and the PSD can be written as

$$|S(f)|^2 \approx \text{rect}\left(\frac{t_0}{T_r}\right)\frac{2\pi}{\theta''(t_0)} \tag{2.32}$$

According the POSP, the phase $\theta(t)$ can be obtained with given PSD $P(f)$ as

$$P(f) = \frac{2\pi}{\theta''(t_0)} \tag{2.33}$$

Further, one can have

$$|S(f)|^2 \approx P(f)|G(f)|^2/2 \tag{2.34}$$

That is to say, the PSD $|S(f)|^2$ of the NLFM waveform consists of the PSD $P(f)$ and the Fresnel integral term $|G(f)|^2$. In fact, since the Fresnel integrals

$$K(\infty) = \sqrt{\frac{j}{2}} \tag{2.35}$$

it follows that

$$|G(0)|^2 \approx 2 \quad |G(-B_r/2)|^2 = |G(B_r/2)|^2 \approx \frac{1}{2} \tag{2.36}$$

Therefore, at the half bandwidth point, the PSD of the NLFM waveform is distorted with respect to the desired weighting PSD due to (2.36). By explicitly incorporating

Table 2.2 Distortion effects for the window function

Weighting function	Δ IRW (sample)	Δ PSLR (dB)
Rectangular[1]	0	-7.1×10^{-5}
Hamming[1]	0.0015	−0.0013
Hanning[1]	0	0
Taylor[1]$(n = 4, sll = -30)$	0	0.125
Chebyshev[1]$(sll = -19)$	0.103	−0.833
Rectangular[2]	0	-8.7×10^{-5}
Hamming[2]	0.0015	−0.0017
Hanning[2]	0	0
Taylor[2]$(n = 4, sll = -30)$	0	0.0833
Dolph[2]$(sll = -19)$	0.064	−0.945

$[\cdot]^1$,$[\cdot]^2$ represents the 50% and 25% edge distortion of window function respectively

this edge distortion into the weighting functions, its influence can be evaluated, as summarized in Table 2.2.

As shown in Table 2.2, compared with the other weighting functions, the mainlobe of the Chebyshev weighting function is clearly broadened by the edge distortion. Because the edge region of the Chebyshev weighting function contains a significant portion of the total energy, the NLFM waveform generated with the Chebyshev window is strongly affected by this edge distortion. To compensate for this effect, a modified Chebyshev weighting for NLFM waveform design is introduced below.

(1). Generate an N-point Chebyshev window for a prescribed peak sidelobe level PSLR_0.

$$\boldsymbol{w} = [w_1, w_2 \ldots, w_N] \tag{2.37}$$

(2). Calculate the IRW of this Chebyshev window and denote it by IRW_0.

(3). Pre-compensate the edge distortion by introducing a weight-modification factor into the Chebyshev window. The resulting modified window is given by

$$\boldsymbol{w}_{\text{mcheb}} = [\alpha w_1, w_2 \ldots, w_{N-1}, \alpha w_N] \tag{2.38}$$

where α is the modification factor satisfying $1 < \alpha < 3$.

(4). For a specified bandwidth B_r and pulse duration T_r, compute the instantaneous frequency function $f_0(t)$ of the NLFM waveform using the modified window $\boldsymbol{w}_{\text{mcheb}}$ based on the POSP procedure.

(5). Increase the linear FM component contained in the instantaneous frequency function. First, compute the linear FM portion of $f_0(t)$ over the intervals $[1, N_t]$ and $[N - N_t + 1, N]$, and determine the corresponding linear FM rate K_p. The linear FM rate itself is kept unchanged. Then, increase and replace the linear FM segment from N_t to $N_t + N_i$ $(N_i = 1, 2 \ldots, N_t)$ so as to obtain an updated instantaneous frequency function $f(t)$.

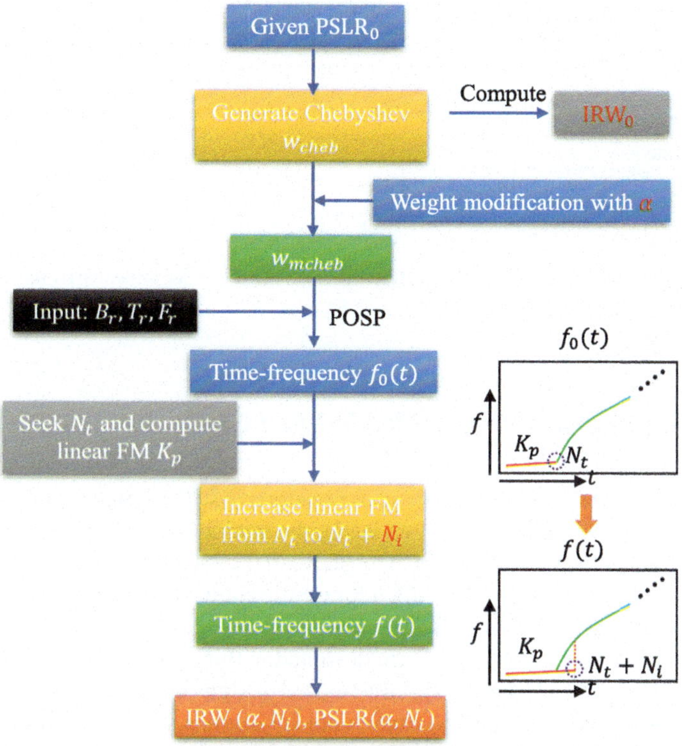

Fig. 2.5 Flowchart for generating NLFM waveform with Mcheb window PSD

(6). For each (α, N_i), compute the IRW and PSLR of the resulting NLFM waveform.

(7). Jointly optimize $\text{IRW}(\alpha, N_i)$ and $\text{PSLR}(\alpha, N_i)$ of the NLFM waveform according to

$$\begin{aligned} &\min_{\alpha, N_i} \text{PSLR}(\alpha, N_i) \\ &\text{s.t. } \text{IRW}(\alpha, N_i) \le \text{IRW}_0 \end{aligned} \tag{2.39}$$

From an optimization perspective, the problem in (2.39) is nonpolynomial-hard (NP-hard). Therefore, we resort to a direct search for a suboptimal solution over the discrete set $(\alpha, N_i)|\alpha = 2 + 0.01n,\ 1 < n < 200,\ 1 \le N_i \le N_t, n \in Z, N_i \in Z$ (where Z denotes the set of integers).

Following the above procedure, a suboptimal solution $(\hat{\alpha}, \hat{N}_i)$ for the NLFM waveform design can be obtained, and the corresponding flowchart for generating the NLFM waveform based on the modified Chebyshev weighting function is depicted in Fig. 2.5.

Table 2.3 Simulation parameters

Parameter	Value	Unit
Pulse width	20	μs
Bandwidth	300	MHz
Sampling rate	360	MHz

The flowchart illustrating the generation of the NLFM waveform using the modified window PSD is presented in Fig. 2.5.

2.3.2 Simulated NLFM with Modified Chebyshev

Table 2.3 summarizes the simulation parameters. As illustrated in Fig. 2.6 (with a zoomed view of the mainlobe in Fig. 2.6d), the autocorrelation of the NLFM waveform employing the MCheb PSD (Chebyshev, $sll = -19.55$ dB, $\alpha = 1.984$) exhibits a peak sidelobe level of −19.57 dB, representing a 6.31 dB improvement over the original LFM signal, and this is achieved without any mainlobe broadening. By comparison, the PSL-optimized waveform designed within the CPM framework attains a PSLR of −18.7 dB [17]. However, CPM-based designs entail high algorithmic complexity, making them difficult to implement for signals with large time–bandwidth products, whereas the proposed modified NLFM waveform can be efficiently realized using the POSP approach.

Figure 2.7 compares the NLFM waveforms obtained with the Chebyshev PSD and the MCheb PSD under the condition of equal IRW. The autocorrelation of the NLFM waveform using the MCheb PSD (Chebyshev, $sll = -39.17$dB, $\alpha = 2.11$) exhibits a peak sidelobe level of −39.51 dB, which represents a 5.09 dB improvement over the Chebyshev ($sll = -38$dB) PSD case, whose peak sidelobe level is −34.43 dB. It is evident that the correlation function of the NLFM waveform designed with the Chebyshev PSD deviates significantly from the sidelobe level associated with the Chebyshev window, whereas the proposed MCheb-based waveform attains sidelobe levels as low as those of the Chebyshev window.

Figure 2.8 depicts the variation of PSLR as a function of the normalized mainlobe width of the NLFM waveform autocorrelation, referenced to the mainlobe width of the LFM signal. The results indicate that, among the three windows considered, the MCheb window offers a markedly superior performance for NLFM waveform design.

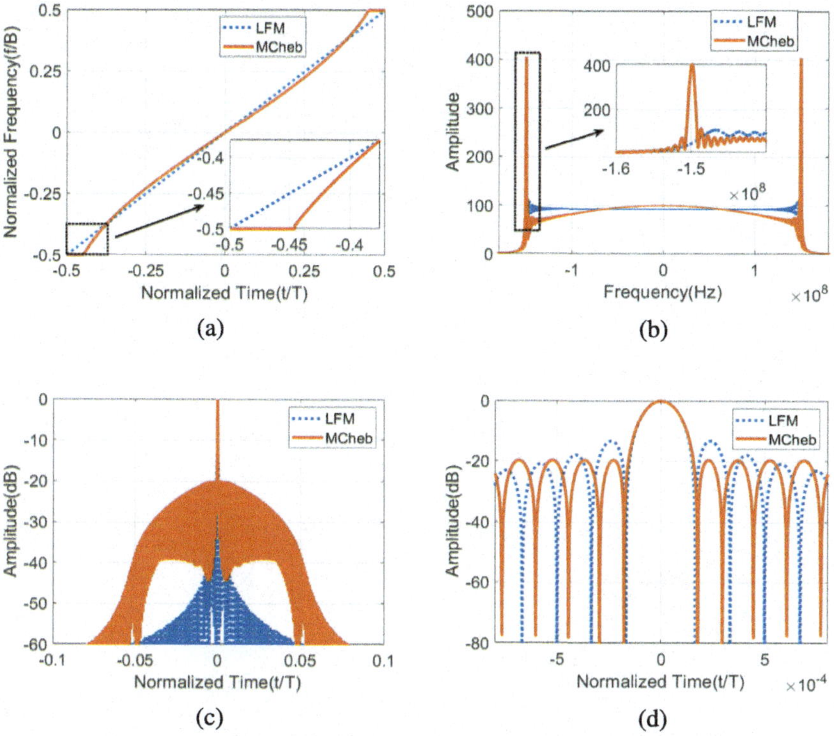

Fig. 2.6 NLFM waveform with MCheb (Chebyshev $(sll = -19.55\text{dB}), \alpha = 1.984$). **a** Time-frequency structure of the NLFM waveform. **b** Spectrum of the NLFM waveform. **c** Autocorrelation of the NLFM waveform. **d** Autocorrelation of the NLFM waveform (close up)

2.4 Real-Time Generator for NLFM Waveform

In this section, we introduce a modified NLFM generation method derived from the principle of stationary phase (POSP) and, on this basis, develop a real-time NLFM signal generator. Furthermore, a two-step compensation strategy is employed to correct system errors, and its performance is verified through ground experiments.

2.4.1 Algorithm Description

In this section, an advanced NLFM generation method is presented that can accurately synthesize NLFM signals with favorable performance while markedly reducing the

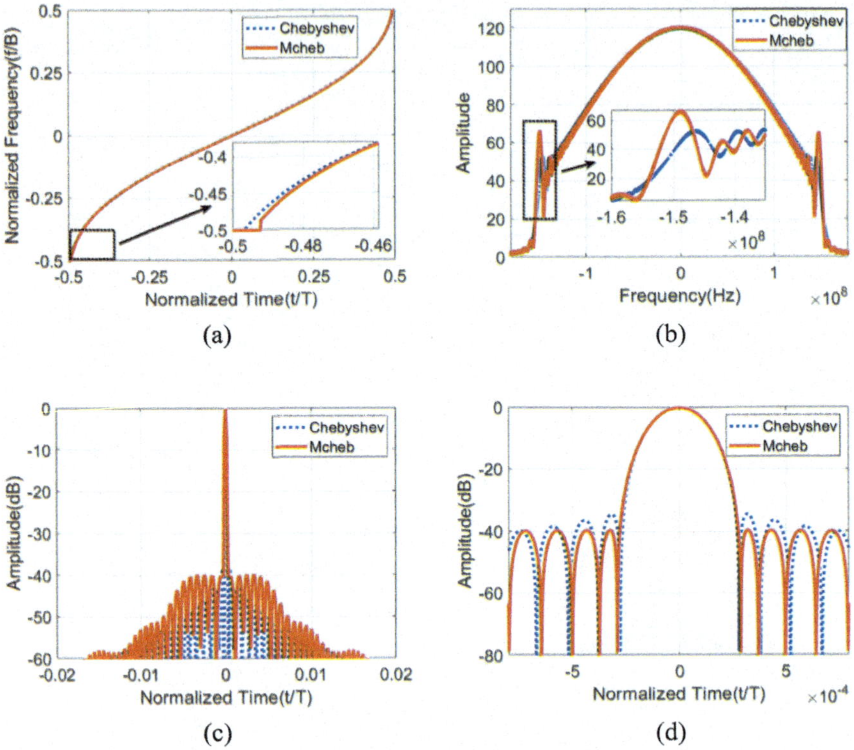

Fig. 2.7 NLFM waveform with MCheb (Chebyshev ($sll = -39.17$dB),$\alpha = 2.11$) and Chebyshev ($sll = -38$dB). **a** Time-frequency structure of the NLFM waveform. **b** Spectrum of the NLFM waveform. **c** Autocorrelation of the NLFM waveform. **d** Autocorrelation of the NLFM waveform (close up)

Fig. 2.8 PSLR to IRW for NLFM with different windows relative to mainlobe width of LFM signal

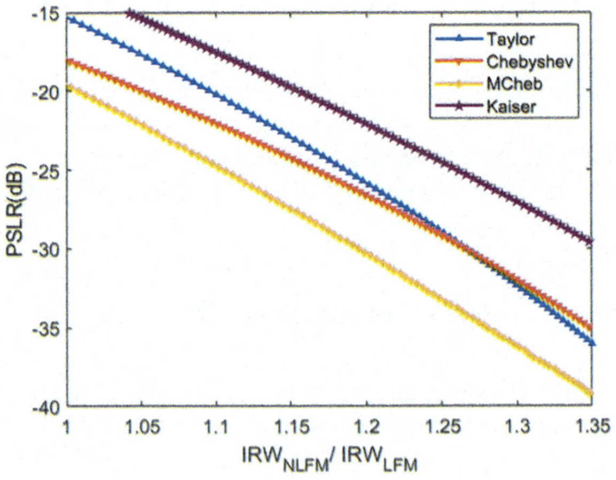

complexity of real-time implementation. Moreover, algorithm complexity analysis and simulation experiments are conducted to verify the effectiveness of the proposed method.

According to (2.8) and (2.9), the instantaneous frequency function $f(t)$ is the inverse of $G(f)$

$$f(t) = G^{-1}(f). \tag{2.40}$$

Then, using the derived $f(t)$, the NLFM signal can be represented as

$$s(t) = \mathrm{rect}\left(\frac{t}{T_r}\right)\exp\left(j2\pi\left(\int_0^t f(\tau)d\tau\right)\right). \tag{2.41}$$

In most cases, an analytic expression for the group time-delay function $G(f)$ can be derived. However, obtaining a closed-form analytic expression for its inverse, namely $f(t)$, is generally difficult. Since $f(t)$ is typically continuous and differentiable, it can be expanded in a polynomial series as

$$f(t) = a_0 + a_1 t + a_2 t^2 + \cdots + a_K t^K + R_K(t) \tag{2.42}$$

where K denotes the polynomial order, a_i $(i = 0, 1, 2, \ldots, K)$ are the polynomial coefficients, and $R_K(t)$ represents the remainder term. To obtain an approximate analytic form, a straightforward approach is to expand $G(f)$ as a polynomial in the frequency f, and then apply the method of series reversion (MSR) to determine the corresponding polynomial coefficients. This procedure, however, is rather cumbersome. In practice, polynomial curve-fitting techniques are more commonly used to approximate the accurate discrete values [10].

For a signal with N sampling points, (2.42) can be rewritten as

$$\boldsymbol{f} = \boldsymbol{G} \cdot \boldsymbol{a} \tag{2.43}$$

where $R_K(t)$ is ignored. And $\boldsymbol{f}$ denotes the discrete frequency vector, as

$$\boldsymbol{f} = [f_0, f_1, \ldots, f_{N-1}]^{\mathrm{T}} \tag{2.44}$$

where $f_i = -\frac{Fs}{2} + i \times \frac{F_s}{N}, i = 0, 1, \ldots, N-1$, F_s is the sampling rate and $[\cdot]^{\mathrm{T}}$ denotes the transpose. And $\boldsymbol{G}$ represents the group time delay matrix, as

$$\boldsymbol{G} = \begin{bmatrix} 1 & g_0 & g_0^2 & \cdots & g_0^K \\ 1 & g_1 & g_1^2 & \cdots & g_1^K \\ 1 & \vdots & \vdots & \ddots & \vdots \\ 1 & g_{N-1} & g_{N-1}^2 & \cdots & g_{N-1}^K \end{bmatrix} \tag{2.45}$$

where $\boldsymbol{g} = (g_0, g_1, \ldots, g_{N-1})$ represents the discrete values of ground time delay which can be obtained by (2.49). Let vector $\boldsymbol{a} = (a_0, a_1, \ldots, a_K)^{\mathrm{T}}$ denote the poly-

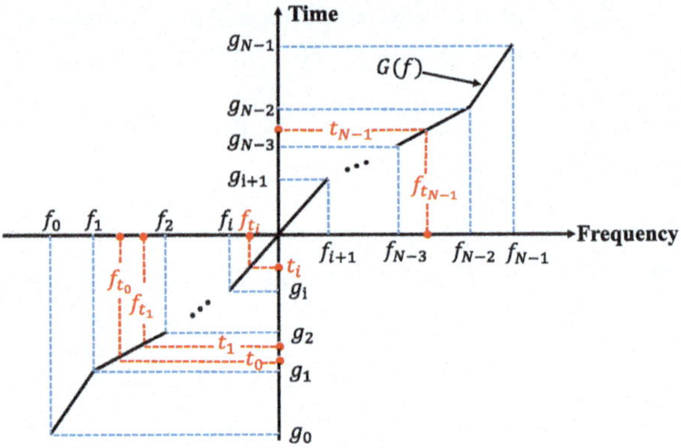

Fig. 2.9 Illustration of the principle of obtaining the instantaneous frequency. Here, $(t_0, t_1, \ldots, t_{N-1})$ denotes the discrete time vector and $(f_{t_0}, f_{t_1}, \ldots, f_{t_{N-1}})$ is the instantaneous frequency vector

nomial coefficient vector. In typical applications, $K \ll N$, so the system is overdetermined and can be formulated as a generalized matrix inversion problem [28]. Accordingly, the coefficient vector is obtained as $\boldsymbol{a} = \boldsymbol{G}^{+}\boldsymbol{f} = (\boldsymbol{G}^T\boldsymbol{G})^{-1}\boldsymbol{G}^T\boldsymbol{f}$. However, implementing matrix inversion on an FPGA is nontrivial, as it involves extensive matrix operations (such as decomposition and multiplication) that demand substantial computational resources and memory [29, 30]. The computational complexity of transposing an $N \times K$ matrix is $O(K \cdot N)$, while the complexity of a matrix product $\boldsymbol{C} = \boldsymbol{A} \cdot \boldsymbol{B}$ is $O(K \cdot N \cdot P)$, where $\boldsymbol{A}$ is a $K \times N$ matrix and $\boldsymbol{B}$ is an $N \times P$ matrix. In addition, for a simple matrix inversion algorithm applied to an $N \times N$ matrix, such as Gauss–Jordan (GJ) elimination, the complexity is on the order of $O(N^3)$ [29]. Consequently, the overall computational burden can be expressed as

$$O(K \cdot N) \times 3 + O(K^2) + O(N \cdot K^2) + O(K^3). \tag{2.46}$$

Generalized piecewise linear (PWL) functions are widely employed to represent complicated nonlinear functions, and the group time-delay function can be approximated in this manner. Specifically, using N sampled points of $G(f)$, the function can be partitioned into $N-1$ linear segments. The corresponding discrete values of the instantaneous frequency $f(t)$ can then be obtained via a straightforward inverse linear interpolation, yielding an algorithmic complexity of $O(N)$. Compared with the polynomial fitting approach, this leads to a substantial reduction in computational complexity. This process is illustrated schematically in Fig. 2.9. Furthermore, for the generation of an NLFM waveform based on a Taylor weighting function with a −40 dB sidelobe, Fig. 2.10a presents the phase difference between the proposed linear interpolation method and the polynomial curve fitting method, while Fig. 2.10b shows the corresponding comparison of the matched-filter (MF) outputs. The simulation results indicate that the performance of the proposed signal generation scheme is very close to that of the polynomial curve fitting method.

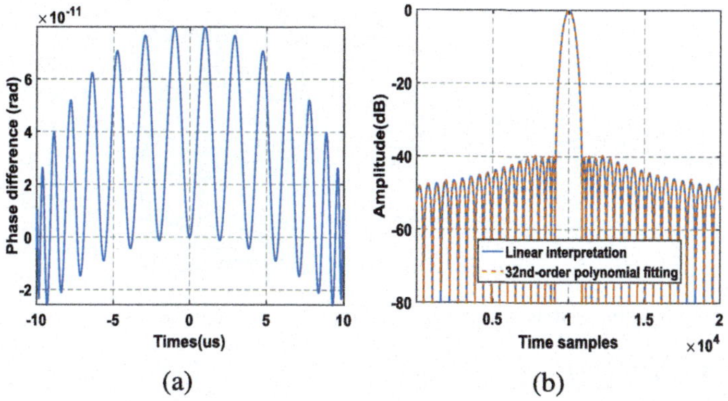

Fig. 2.10 Performance comparison between the linear interpolation and polynomial curve fitting ($B = 150$MHz, $T_p = 20\,\mu$s). **a** Phase difference. **b** MF output

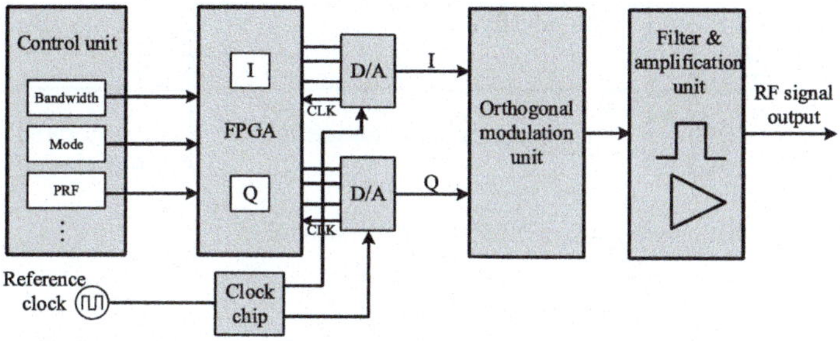

Fig. 2.11 Simplified block diagram of the signal generator

2.4.2 *Hardware Architecture*

In this section, the implementation of the NLFM signal generator, developed on the basis of the waveform-generation method described in the previous section, is presented. A simplified block diagram of the generator is shown in Fig. 2.11. After receiving system parameters (such as bandwidth, imaging mode, PRF, etc.) via the serial port, the control unit configures the FPGA to produce the corresponding signal. A low-phase-noise reference clock, required to satisfy the ADC phase-noise specifications given by the manufacturer, is generated by a dedicated clock chip. The digital waveform computed in the FPGA is then converted into the corresponding analog signal by two ADC chips, after which the baseband signal is upconverted to the radio-frequency (RF) signal by an orthogonal modulation unit. Finally, the RF signal is filtered, amplified, and delivered to various radar modules.

To further clarify the computation principle of the developed NLFM generator, Fig. 2.12 presents a simplified block diagram of the processing flow implemented in the FPGA. Upon receiving configuration instructions from the host computer via the serial port, the signal parameters, including bandwidth, pulse width, and

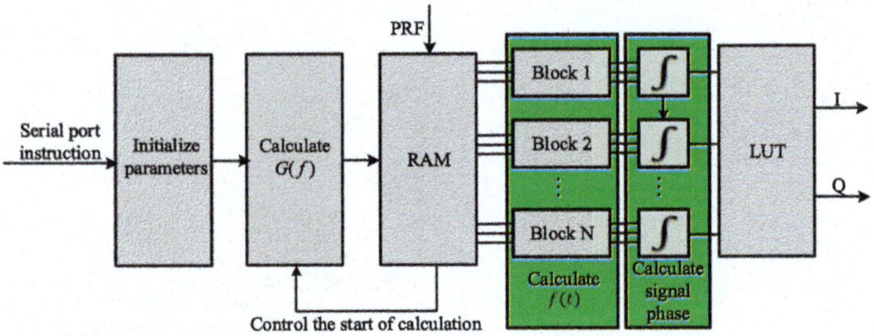

Fig. 2.12 Simplified block diagram of the calculation process in the FPGA

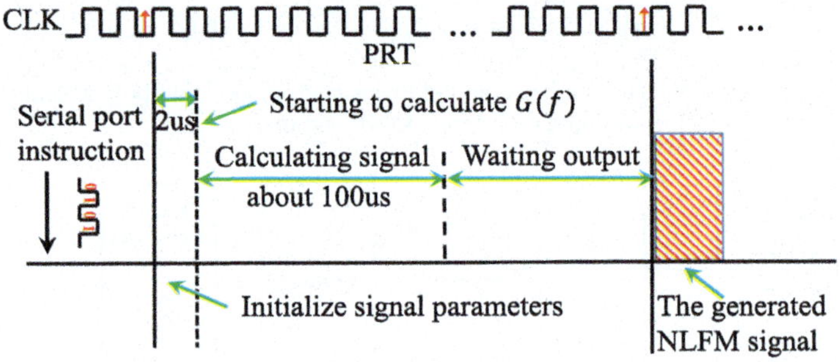

Fig. 2.13 The calculation sequence of NLFM signal generator

PSLR, are initialized. The group time function is then evaluated using a sin look-up table (LUT). It should be emphasized that $G(f)$ is not computed directly from (2.20); instead, appropriate scaling is applied to ensure numerical accuracy. After obtaining the group time function, the data are partitioned into multiple blocks, and the instantaneous frequency in each block is computed in parallel by simple linear interpolation. Subsequently, the signal phase in each block is obtained by integrating the corresponding instantaneous-frequency samples, where the final phase value of the previous block is used as the initial phase of the next block, thereby completing the phase calculation of the NLFM signal. Finally, the discrete NLFM waveform samples are generated by means of a sin-cos LUT.

Moreover, the timing sequence of the entire computation is depicted in Fig. 2.13. After the serial-port command is received, approximately 2 μs are required to initialize the signal parameters. The FPGA then spends about 100 μs performing the calculation steps outlined in Fig. 2.12. The corresponding NLFM signal is subsequently output in the next PRT. A photograph of the developed NLFM signal generator is provided in Fig. 2.14.

Fig. 2.14 Photograph of the developed NLFM signal generator

2.4.3 Experimental Result

In this section, we present the experimental results and introduce a two-step error-compensation scheme to correct waveform distortions. The RF NLFM signals produced by the signal source under different parameter settings are sampled by an oscilloscope, after which a digital orthogonal demodulation procedure is applied to recover the corresponding baseband signals. The performance comparison between the FPGA-generated signals and the ideal reference waveforms is shown in Fig. 2.15. It can be observed that, in all cases, the characteristics of the generated signals closely match the theoretical ones, confirming that the developed signal generator achieves high computational accuracy.

Furthermore, a ground experiment is conducted. As an example, when the system operates with a signal (predefined PSLR: −40 dB, $B = 60$ MHz, $T_p = 80\,\mu$s) without predistortion, the compressed result of the received baseband signal is obtained, as shown in Fig. 2.16a. In the first error-compensation step, the baseband signal is modified to mitigate the link-induced errors. By averaging the phase over multiple pulses to reduce the influence of random noise, the phase error is expressed as

$$\bar{\phi}_d(t) = \left[\frac{1}{M}\sum_{i=1}^{M}\mathbf{phase}(s_{d,i}(t))\right] - \phi(t) \tag{2.47}$$

where $s_{d,i}(t)$ denotes the ith received pulse, M is the number of pulses, and $\phi(t)$ is the phase of the reference signal. What's more, $\mathbf{phase}(\cdot)$ represents phase unwrapping. The corresponding phase-error curve is presented in Fig. 2.16b, from which it can be observed that the error is dominated by high-order components, leading to pronounced asymmetric distortion. Likewise, the mean amplitude error is evaluated as

$$\bar{a}_d(t) = \frac{1}{M}\sum_{i=1}^{M}|s_{d,i}(t)|. \tag{2.48}$$

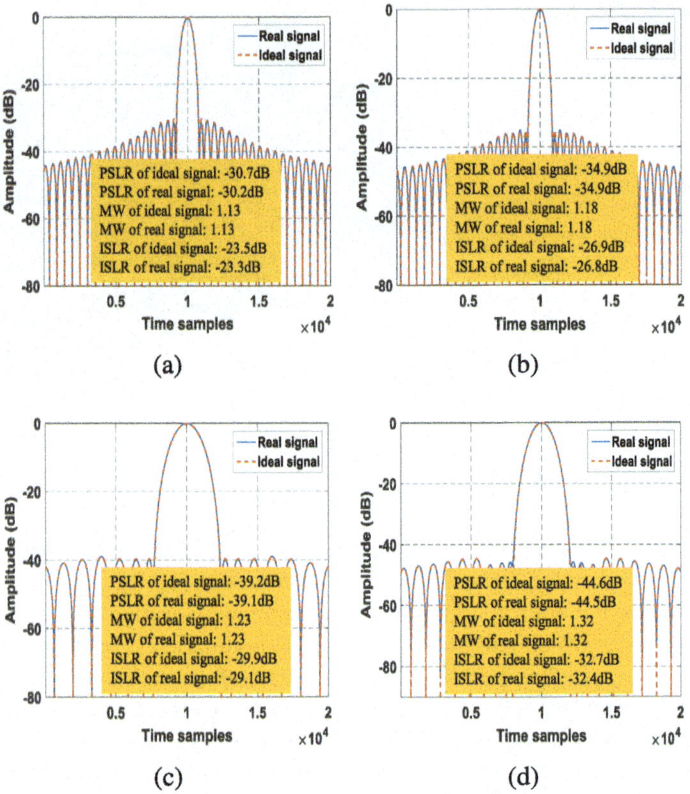

Fig. 2.15 MF output performance comparison between the generated signal and the ideal signal in different signal parameters. **a** Predefined PSLR: −30 dB, $B = 150$ MHz, $T_p = 110\,\mu s$. **b** Predefined PSLR: −35 dB, $B = 150$ MHz, $T_p = 30\,\mu s$. **c** Predefined PSLR: −40 dB, $B = 60$ MHz, $T_p = 80\,\mu s$. **d** Predefined PSLR: −45 dB, $B = 80$ MHz, $T_p = 110\,\mu s$

Theoretically, signals with identical bandwidth and sidelobe levels should experience the same system link errors, irrespective of pulse width. In practice, however, quantization effects give rise to residual high-order phase errors that persist even after phase-error compensation. Therefore, the influence of each quantization stage, such as signal generation and data acquisition, must be taken into account. To extract the system error more accurately, the quantized NLFM signal is adopted as the reference waveform. Owing to the limited ROM capacity, storing all discrete phase-error samples is impractical. As shown in Fig. 2.16b, only a small number of phase-error control points are stored, and these control points are subsequently expanded to the full data length during phase computation in order to compensate the system phase error.

Owing to the full power transmitting, amplitude compensation at baseband is ineffective. The experiment was therefore repeated using the phase-predistorted signal, and the compressed result of the collected data is shown in Fig. 2.17a. It can be

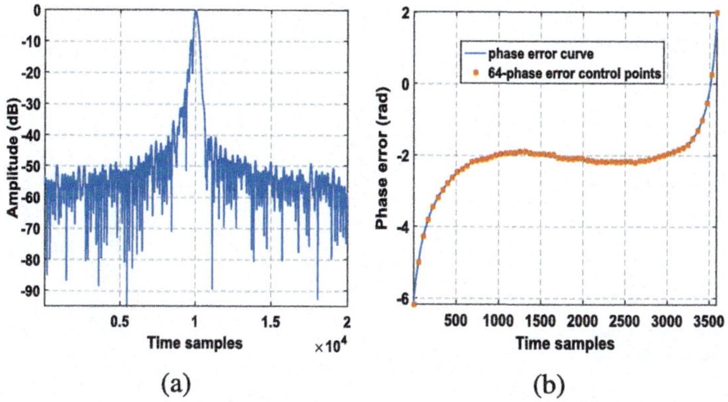

Fig. 2.16 Example of recorded baseband signal ($B = 60\text{MHz}$, $T_p = 80\,\mu\text{s}$). **a** Compressed result of the distorted signal. **b** Phase error

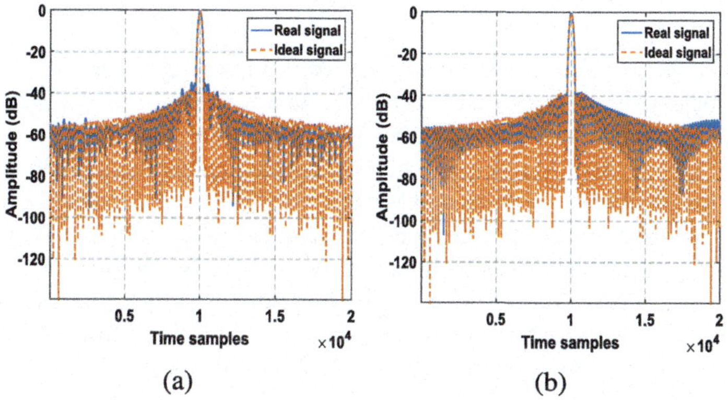

Fig. 2.17 Compression results of the received signal after compensation. **a** First compensation result. **b** Second compensation result

seen that, after phase predistortion, the asymmetric distortion disappears; however, the PSLR remains −34.5 dB, which is still higher than the theoretical value. When the system operates on orbit, its state (e.g., temperature) varies over time. Thus, a calibration signal can be further employed during the imaging process to compensate the residual phase-amplitude errors, constituting a second error-compensation step. After this second compensation step, the compressed result shown in Fig. 2.17b is obtained. The compressed signal exhibits a peak sidelobe level of −38.5 dB, which is close to the theoretical value in Fig. 2.15c, thereby validating the effectiveness of the proposed two-step compensation method.

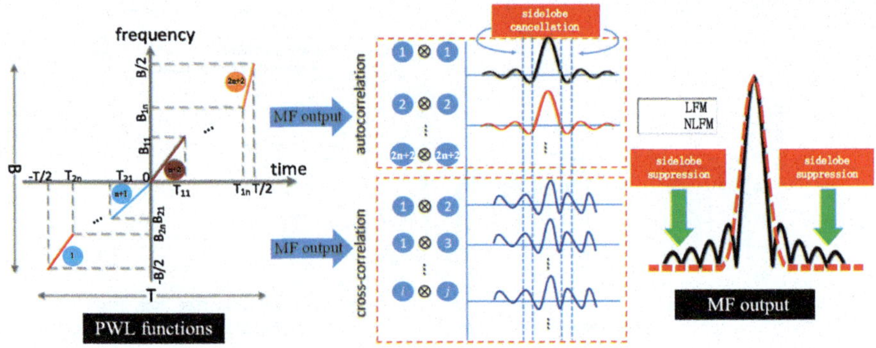

Fig. 2.18 $2n + 2$ stages for PWL functions

2.5 NLFM Waveform Optimization Method

Cook and Paolillo demonstrated that a waveform employing linear FM over most of the pulse, but incorporating short segments with a higher FM rate at the beginning and end, can effectively reduce sidelobes [31]. They attributed this improvement to the ability of the higher FM rate segments to suppress Fresnel ripples in the waveform spectrum. Subsequently, Griffiths and Vinagre [32] offered an alternative interpretation of this technique using a piecewise NLFM waveform, as illustrated in Fig. 2.18, arguing that, with an appropriate choice of the higher FM rate segments, the sidelobes of the autocorrelation and cross-correlation functions of each segment mutually cancel, thereby lowering the overall sidelobe level. Motivated by this concept, we employ PWL functions to parameterize the NLFM waveform.

2.5.1 *NLFM Waveform Descriptions*

An instantaneous frequency function is constructed using $2n + 2$ stages in the PWL representation, where $2n$ denotes the number of frequency control points, as illustrated in Fig. 2.18. Unlike [16], the optimal instantaneous frequency function is not assumed to be symmetric, and an asymmetric PWL formulation is therefore adopted. The instantaneous frequency of the proposed NLFM waveform can be expressed as follows:

$$f_{+}(t) = \begin{cases} k_{10}t & 0 \le t < T_{11} \\ B_{11} + k_{11}(t - T_{11}) & T_{11} \le t < T_{12} \\ \cdots & \cdots \\ B_{1n} + k_{1n}(t - T_{1n}) & T_{1n} \le t < T/2 \end{cases} \quad (2.49)$$

$$f_-(t) = \begin{cases} k_{20}t & T_{21} \le t < 0 \\ B_{21} + k_{21}(t - T_{21}) & T_{22} \le t < T_{21} \\ \cdots & \cdots \\ B_{2n} + k_{2n}(t - T_{2n}) & -T/2 \le t < T_{2n} \end{cases} \tag{2.50}$$

where k_{1i} and k_{2i} represent the FM rate of each portions as follows:

$$k_{10} = \frac{B_{11}}{T_{11}}, \ldots, \underset{1 \le i < n}{k_{1i}} = \frac{B_{1(i+1)} - B_{1i}}{T_{1(i+1)} - T_{1i}}, \ldots, k_{1n} = \frac{\frac{B}{2} - B_{1n}}{\frac{T}{2} - T_{1n}}. \tag{2.51}$$

$$k_{20} = \frac{B_{21}}{T_{21}}, \ldots, \underset{1 \le i < n}{k_{2i}} = \frac{B_{2(i+1)} - B_{2i}}{T_{2(i+1)} - T_{2i}}, \ldots, k_{2n} = \frac{\frac{B}{2} + B_{2n}}{\frac{T}{2} + T_{2n}}. \tag{2.52}$$

Then, the phase of this 2n + 2 stages NLFM waveform can be derived by integrating (2.49) and (2.50) as

$$\theta_+(t) = 2\pi \begin{cases} k_{10}t^2 + c_0 & 0 \le t < T_{11} \\ B_{11}t + k_{11}(t^2/2 - T_{11}t) + c_1 & T_{11} \le t < T_{12} \\ \cdots & \cdots \\ B_{1n} + k_{1n}(t^2/2 - T_{1n}t) + c_n & T_{1n} \le t < T/2 \end{cases} \tag{2.53}$$

$$\theta_-(t) = 2\pi \begin{cases} k_{20}t^2 + d_0 & T_{21} \le t < 0 \\ B_{21}t + k_{21}(t^2/2 - T_{21}t) + d_1 & T_{22} \le t < T_{21} \\ \cdots & \cdots \\ B_{2n} + k_{2n}(t^2/2 - T_{2n}t) + d_n & -T/2 \le t < T_{2n} \end{cases} \tag{2.54}$$

where c_i and d_i are constant

$$c_0 = 0, c_1 = -\frac{1}{2}B_{11}T_{11} + \frac{1}{2}k_{11}T_{11}^2 \tag{2.55}$$

$$\begin{aligned} \underset{2 \le i < n}{c_i} &= (B_{1(i-1)} - B_{1i})T_{1i} + \frac{1}{2}T_{1i}^2(k_{1(i-1)} + k_{1i}) \\ &\quad - k_{1(i-1)}T_{1i}T_{1(i-1)} + c_{i-1} \end{aligned} \tag{2.56}$$

$$d_0 = 0, d_1 = -\frac{1}{2}B_{21}T_{21} + \frac{1}{2}k_{21}T_{21}^2 \tag{2.57}$$

$$\begin{aligned} \underset{2 \le i < n}{d_i} &= (B_{2(i-1)} - B_{2i})T_{2i} + \frac{1}{2}T_{2i}^2(k_{2(i-1)} + k_{2i}) \\ &\quad - k_{2(i-1)}T_{2i}T_{2(i-1)} + d_{i-1} \end{aligned} \tag{2.58}$$

2.5.2 Augmented Lagrangian Genetic Algorithm

By enforcing equal duration for all segments, we set $T_{1i} = \frac{T}{2n+2}i, T_{2j} = -\frac{T}{2n+2}i, (i = 1, 2, 3, \ldots, n)$, so that there are $2n$ design parameters ($B_{11} \ldots B_{1n}$ and $B_{21} \ldots B_{2n}$) to be optimized in order to obtain the desired NLFM waveform. The performance of the resulting NLFM waveform is evaluated in terms of PSLR and IRW.

To design an NLFM waveform that achieves sidelobe reduction without broadening the mainlobe, we formulate the task as a nonlinear constrained optimization problem:

$$\begin{aligned}
&\min_x f(x) \text{ such that} \\
&c_i(x) \le 0, i = 1, \ldots, m \\
&ceq_i(x) = 0, i = m+1, \ldots, M \\
&A * x \le b, Aeq * x = beq, lb \le x \le ub.
\end{aligned} \tag{2.59}$$

where $c(x)$ denotes the nonlinear inequality constraints, $ceq(x)$ denotes the nonlinear equality constraints, m is the number of nonlinear inequality constraints, and M is the total number of nonlinear constraints. Accordingly, the PSLR-oriented optimization problem can be written as

$$\begin{aligned}
&\min_{\boldsymbol{B}} f(\text{PSLR}), \text{ such that} \\
&c(\text{IRW}) \le 0, -B/2 \le \boldsymbol{B} \le B/2,
\end{aligned} \tag{2.60}$$

where $\boldsymbol{B} = (B_{11} \ldots B_{1n}, B_{21} \ldots B_{2n})$ collects all the frequency-control parameters to be optimized.

The Genetic Algorithm (GA) is an adaptive heuristic search technique inspired by the principles of population genetics. Since its introduction by John Holland in the early 1970s [33], GA has been extensively applied to global optimization problems. The Augmented Lagrangian Genetic Algorithm (ALGA) [34] extends the conventional GA framework to handle nonlinear constrained optimization problems of the form

$$\begin{aligned}
\Theta(x, \lambda, s, \rho) = f(x) - \sum_{i=1}^{m} \lambda_i s_i \log(s_i - c_i(x)) \\
+ \sum_{i=m+1}^{mt} \lambda_i ceq_i(x) + \frac{\rho}{2} \sum_{mt}^{i=m+1} ceq_i(x)^2.
\end{aligned} \tag{2.61}$$

Here, $f(x)$ denotes the fitness function; the components λ_i of the vector λ are nonnegative Lagrange multiplier estimates; the elements s_i of the vector s are nonnegative shift parameters; and ρ is a positive penalty parameter. For the PSLR-oriented optimization problem, the augmented Lagrangian can be written as

$$\Theta(\boldsymbol{B}, \lambda, s) = f(\boldsymbol{B}) - \lambda s \log(s - c(\boldsymbol{B})) \tag{2.62}$$

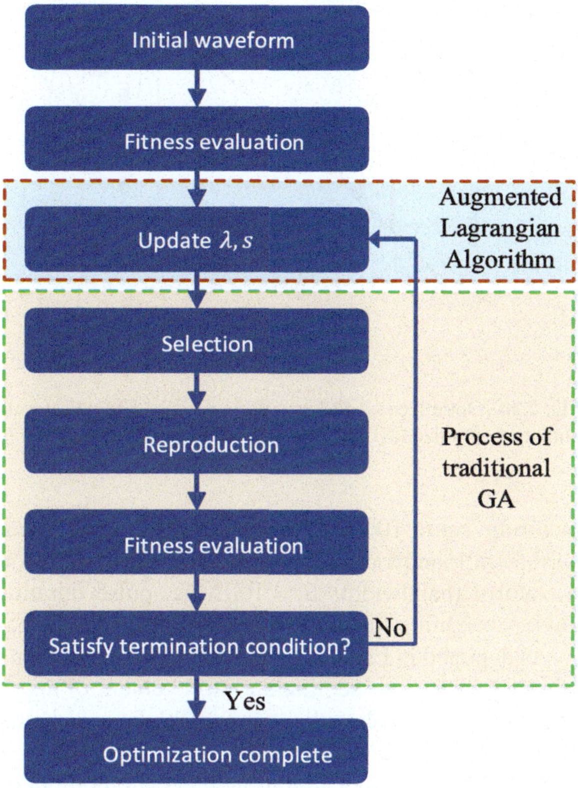

Fig. 2.19 Flowchart for the NLFM waveform optimization

The ALGA proceeds by successively minimizing a series of subproblems, where the solution of each subproblem corresponds to one generation and serves as an approximation to the original constrained optimization problem. Each subproblem is constructed by augmenting the fitness function with the nonlinear constraint functions through Lagrangian terms and associated penalty parameters. When a subproblem has been solved to a prescribed accuracy and the feasibility conditions are satisfied, the estimates of the Lagrange multipliers are updated; otherwise, the penalty parameter is increased by a given factor. The main stages of the NLFM waveform optimization procedure are summarized in Fig. 2.19. A detailed exposition of the ALGA mechanism is beyond the scope of this paper, and interested readers are referred to [34] for further information.

2.5.3 Simulation Results

The following numerical examples are obtained using a 130-stage PWL representation ($n = 64$). For the ALGA, the population size, crossover factor, and mutation

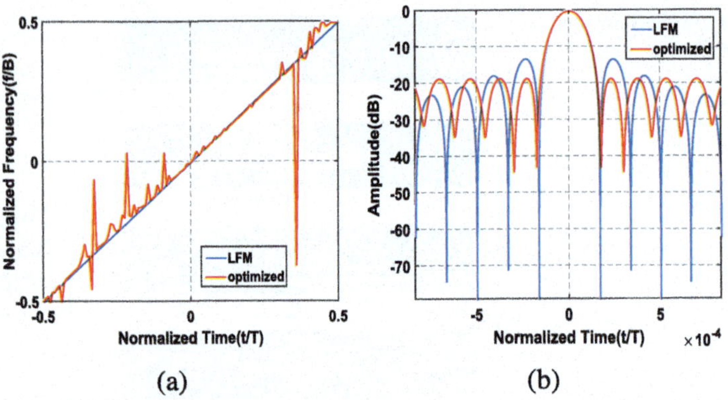

Fig. 2.20 Optimized NLFM waveform versus LFM waveform. **a** Time-frequency structure. **b** MF output of optimized NLFM waveform versus LFM waveform (close up)

factor are set to 100, 0.8, and 0.5, respectively. The initial waveforms are an LFM waveform (bandwidth: $B = 300$ Mhz, pulse duration: $T = 20$ μs) and an NLFM waveform (bandwidth: $B = 100$ Mhz, pulse duration: $T = 10$ μs) generated by a Taylor weighting function with a −35 dB sidelobe based on the POSP.

As depicted in Fig. 2.20b, the optimized waveform achieves a peak sidelobe level of −18.51 dB, representing an improvement of approximately 5 dB relative to the original LFM signal, without any mainlobe broadening. The corresponding normalized time–frequency distribution is shown in Fig. 2.20a. It is noteworthy that this structure contains short-duration wideband components reminiscent of the echolocation emissions of certain bat species [35, 36], a feature also reported in [17, 19].

Another example is presented in Fig. 2.21, with a magnified view of the mainlobes of the different NLFM waveforms shown in Fig. 2.22. The corresponding performance comparison is summarized in Table 2.4. The optimized waveform attains a peak sidelobe level of −40.2 dB, yielding an improvement of 6 dB over the original NLFM signal while preserving the mainlobe width. In addition, relative to the NLFM waveform designed using a −42 dB Taylor weighting function, the proposed waveform reduces the normalized mainlobe width from 1.26 to 1.17 without increasing the sidelobe level. Furthermore, Table 2.4 shows that the performance of the NLFM waveforms proposed in [37, 38] is comparable to that of the POSP-based NLFM design, since all these methods rely on underlying weighting functions and mainly differ in how the PSD is shaped. By contrast, when compared with the optimized NLFM waveforms reported in [15, 16], the proposed NLFM designs exhibit superior performance in terms of both IRW and PSLR.

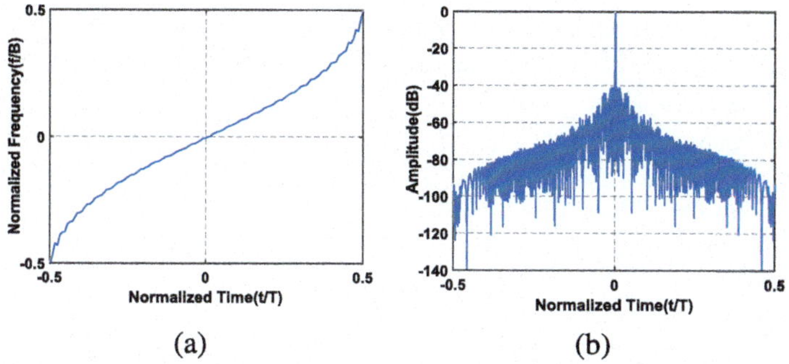

Fig. 2.21 Optimized NLFM waveform. **a** Time-frequency structure. **b** MF output

Fig. 2.22 MF output of the optimized NLFM waveform versus the NLFM waveforms based on Taylor weighting functions (close up)

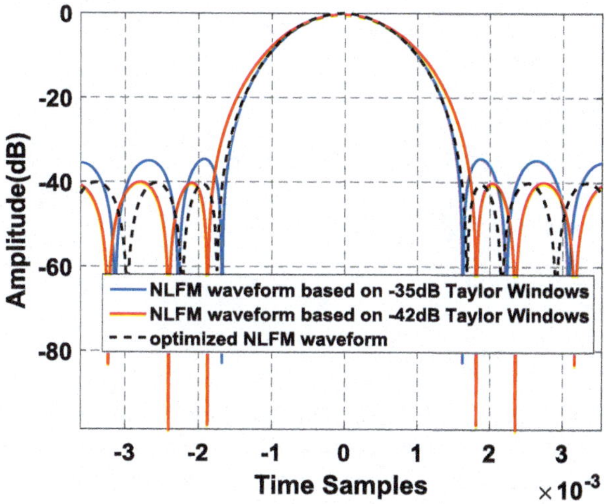

2.6 Summary

NLFM signals can suppress sidelobes without incurring SNR loss and are therefore attractive alternatives to LFM signals. However, (i) they frequently lead to mainlobe broadening, and (ii) the development of efficient and accurate waveform-generation algorithms suitable for hardware implementation remains challenging. To tackle these issues, this chapter first presents a novel NLFM waveform based on a modified Chebyshev weighting function, which achieves low PSLR while maintaining the lowest sidelobe level inherent to the Chebyshev weighting function. Simulation results show that the NLFM waveform with the modified Chebyshev weighting not only attains Chebyshev-level sidelobes, but also does so without demanding high computational complexity.

Secondly, a modified NLFM signal generation scheme based on POSP is presented. On this basis, an NLFM signal generator with predistortion compensation

Table 2.4 The Performance comparison of different NLFM waveforms

Waveform	IRW (sample)	PSLR (dB)
LFM waveform	0.89	−13.3
NLFM waveform based on POSP		
—Taylor window($\bar{n} = 4$, $sll = -35$)	1.19	−34.3
—Taylor window($\bar{n} = 4$, $sll = -35$)	1.26	−40
Proposed optimized NLFM waveform		
—Initialized by an LFM waveform	0.89	−18.51
—Initialized by an NLFM waveform	1.17	−40.2
NLFM generation method proposed in [37]	1.26	−40
Cosine, $n = 3$ [6]	1.67	−38.8
Iterative method [38]	1.27	−40
Optimized NLFM waveform [15]	1.2	−35.7
Optimized NLFM waveform, $v = 0.9$ [16]	0.9	−16.96
Hybrid NLFM [12]	1.56	−38.4

capability is developed, which has been implemented in the LT-1 mission to demonstrate the performance of NLFM signalling in spaceborne SAR imaging. In addition, a two-step error-compensation method is proposed, and ground experimental results are provided to validate both the effectiveness of the proposed compensation strategy and the performance of the developed signal generator.

Finally, a novel optimization framework for NLFM waveform design based on ALGA and PWL functions is introduced. In this approach, the instantaneous frequency is parameterized by PWL functions through a set of frequency control points, thereby specifying the waveform over the entire time–frequency plane. Increasing the number of control points enriches the representational flexibility of the waveform, but also enlarges the associated search space. ALGA is employed to carry out the waveform optimization over these parameters. Within this framework, an advanced NLFM waveform with reduced sidelobes and a narrower mainlobe is obtained, and simulation results substantiate the effectiveness of the developed NLFM designs.

References

1. Shayesteh MG, Mottaghi-Kashtiban M. FIR filter design using a new window function. In: 2009 16th International Conference on Digital Signal Processing; 2009. p. 1–6.
2. Ghavamirad R, Babashah H, Sebt MA. Nonlinear FM waveform design to reduction of sidelobe level in autocorrelation function. In: 2017 Iranian Conference on Electrical Engineering (ICEE); 2017. p. 1973–1977.
3. Ghavamirad R, Sebt MA. Sidelobe Level Reduction in ACF of NLFM Waveform. IET Radar, Sonar & Navigation. 2018;.
4. Ghavamirad R, Sebt MA, Babashah H. Phase Improvement Algorithm for NLFM Waveform Design to Reduction of Sidelobe Level in Autocorrelation Function. Electronics Letters. 2018;.
5. Fowle E. The design of FM pulse compression signals. IEEE Transactions on Information Theory. 1964;10(1):61–67.
6. Cook CE. A class of nonlinear FM pulse compression signals. Proceedings of the IEEE. 1964 Nov;52(11):1369–1371.
7. de Buda R. Stationary phase approximations of FM spectra. IEEE Transactions on Information Theory. 1966 July;12(3):305–311.
8. Vizitiu I, Anton L, Popescu F, et al. The synthesis of some NLFM laws using the stationary phase principle. In: 2012 10th International Symposium on Electronics and Telecommunications; 2012. p. 377–380.
9. Zhang Y, Wang W, Wang R, et al. A novel NLFM waveform with low sidelobes based on modified Chebyshev window. IEEE Geoscience and Remote Sensing Letters. 2019;17(5):814–818.
10. Wang W, Wang R, Zhang Z, et al. First Demonstration of Airborne SAR With Nonlinear FM Chirp Waveforms. IEEE Geoscience and Remote Sensing Letters. 2016 Feb;13(2):247–251.
11. Collins T, Atkins P. Nonlinear frequency modulation chirps for active sonar. IEE Proceedings-Radar, Sonar and Navigation. 1999;146(6):312–316.
12. Levanon N, Mozeson E. Radar signals. John Wiley & Sons; 2004.
13. Alphonse S, Williamson GA. Novel radar signal models using nonlinear frequency modulation. In: 2014 22nd European Signal Processing Conference (EUSIPCO). IEEE; 2014. p. 1024–1028.
14. Jin G, Liu K, Deng Y, et al. Nonlinear frequency modulation signal generator in LT-1. IEEE Geoscience and Remote Sensing Letters. 2019;16(10):1570–1574.
15. Kurdzo JM, Cheong BL, Palmer RD, et al. Optimized NLFM pulse compression waveforms for high-sensitivity radar observations. In: 2014 International Radar Conference. IEEE; 2014. p. 1–6.
16. Saeedi J, Faez K. Synthetic aperture radar imaging using nonlinear frequency modulation signal. IEEE Transactions on Aerospace and Electronic Systems. 2016 February;52(1):99–110.
17. Jakabosky J, Anglin P, Cook MR, et al. Non-linear FM waveform design using marginal Fisher's information within the CPM framework. In: Radar Conference (RADAR), 2011 IEEE. IEEE; 2011. p. 513–518.
18. Jakabosky J, Blunt SD, Higgins T. Ultra-low sidelobe waveform design via spectral shaping and LINC transmit architecture. In: 2015 IEEE Radar Conference (RadarCon); 2015. p. 1021–1026.
19. Blunt SD, Jakabosky J, Cook M, et al. Polyphase-coded FM (PCFM) radar waveforms, part II: optimization. IEEE Transactions on Aerospace and Electronic Systems. 2014;50(3):2230–2241.
20. Jin G, Deng Y, Wang R, et al. An Advanced Nonlinear Frequency Modulation Waveform for Radar Imaging With Low Sidelobe. IEEE Transactions on Geoscience and Remote Sensing. 2019;57(8):6155–6168.
21. Papoulis A. Signal analysis. Mcgraw-Hill College; 1977.
22. Lynch P. The Dolph–Chebyshev window: A simple optimal filter. Monthly weather review. 1997;125(4):655–660.

23. Kaiser JF. Nonrecursive digital filter design using the I_0-sinh window function. In: Proc. 1974 IEEE International Symposium on Circuits & Systems, San Francisco DA, April; 1974. p. 20–23.
24. Van Trees HL. Optimum array processing: Part IV of detection, estimation, and modulation theory. John Wiley & Sons; 2004.
25. Hansen EW. Fourier transforms: principles and applications. John Wiley & Sons; 2014.
26. Oppenheim AV. Discrete-time signal processing. Pearson Education India; 1999.
27. Dolph C. A current distribution for broadside arrays which optimizes the relationship between beam width and side-lobe level. Proceedings of the IRE. 1946;34(6):335–348.
28. Tveito A, Langtangen HP, Nielsen BF, et al. Parameter Estimation and Inverse Problems. Academic Press,; 2013.
29. Arias-García J, Jacobi RP, Llanos CH, et al. A suitable FPGA implementation of floating-point matrix inversion based on Gauss-Jordan elimination. In: 2011 VII Southern Conference on Programmable Logic (SPL); 2011. p. 263–268.
30. Arias-García J, Llanos CH, Ayala-Rincón M, et al. FPGA implementation of large-scale matrix inversion using single, double and custom floating-point precision. In: 2012 VIII Southern Conference on Programmable Logic; 2012. p. 1–6.
31. Cook CE, Paolillo J. A pulse compression predistortion function for efficient sidelobe reduction in a high-power radar. Proceedings of the IEEE. 1964 April;52(4):377–389.
32. Griffiths HD, Vinagre L. Design of low-sidelobe pulse compression waveforms. Electronics Letters. 1994 Jun;30(12):1004–1005.
33. Holland JH. Adaption in Natural and Artificial Systems. Quarterly Review of Biology. 1975;6(2):126–137.
34. Conn A, Gould N, Toint P. A globally convergent Lagrangian barrier algorithm for optimization with general inequality constraints and simple bounds. Mathematics of Computation of the American Mathematical Society. 1997;66(217):261–288.
35. Vespe M, Jones G, Baker CJ. Lessons for Radar. IEEE Signal Processing Magazine. 2009 Jan;26(1):65–75.
36. Balleri A, Griffiths H, Holderied M, et al. Bat-inspired multi-harmonic waveforms. In: 2010 International Waveform Diversity and Design Conference; 2010. p. 000086–000089.
37. Doerry AW. Generating nonlinear FM chirp waveforms for radar. Sandia National Laboratories; 2006.
38. Jackson L, Kay S, Vankayalapati N. Iterative Method for Nonlinear FM Synthesis of Radar Signals. IEEE Transactions on Aerospace and Electronic Systems. 2010 April;46(2):910–917.

Chapter 3
The Application of NLFM Waveform in SAR

Along with the advancement of high-resolution radar satellites for global all-round environmental monitoring, round-the-clock and all-weather surveillance has become a vibrant and expanding research area. However, in all cases, these applications demand that radar be equipped with a high-efficiency radar module (e.g., T/R module), and substantial system transmission power. Such requirements might impose a crucial constraint on the performance of a high-power radar satellite. In contrast to the linear frequency modulation (LFM) waveform, the nonlinear frequency modulation (NLFM) waveform exhibits lower sidelobes without sacrificing the signal-to-noise ratio (SNR), which can notably alleviate the constraint of extremely limited satellite power. In this chapter, we employ the NLFM waveform to synthetic aperture radar (SAR) system. Firstly, the expressions of time and frequency matched filter are derived for NLFM waveform. Further, the modified range migration algorithm (RMA) and modified chirp scaling algorithm (CSA) are devised to make them suitable for focusing NLFM-SAR data. In addition, for spaceborne NLFM SAR, the effect of Doppler shift on NLFM SAR imaging cannot be neglected. Therefore, to further advance the application of NLFM in SAR, an effective compensation method is proposed. Finally, we apply the NLFM waveform in an actual synthetic aperture radar (SAR) system with a bandwidth of 100 MHz at a carrier frequency of 9.6 GHz. The imaging results illustrate the outstanding performance of the advanced NLFM waveform. Furthermore, we find that the SAR system utilizing the advanced waveform has a 1.29 dB higher SNR compared to the traditional linear frequency modulation (LFM) waveform. This 1.29 dB SNR improvement implies that the actual radar system can cut down transmitting power by 25%. This characteristic is likely to be a potential advantage of the NLFM waveform, which can reduce the transmitting power demand, particularly for radar satellites.

R. Wang and W. Wang, *Synthetic Aperture Radar (SAR) Waveform Design and Signal Processing*, https://doi.org/10.1007/978-981-95-7549-7_3

3.1 Introduction

Synthetic aperture radar (SAR) is a technology where backscattered microwave pulses, collected by an aircraft or satellite, are processed to create an image of the Earth's surface [1]. The origin of SAR can be traced back to 1951, and presently, numerous new techniques have been developed, such as interferometric SAR (InSAR) [2] and polarimetric SAR (PolSAR) [3]. Typically, a SAR system emits a linear frequency modulation (LFM) waveform (for example, TerraSAR-X (e.g., TerraSAR-X [4, 5]) and pulse compression technology is utilized to enhance the processing gain. However, the SAR system employing the NLFM waveform has a higher signal-to-noise This 1.29 dB SNR improvement means that a real radar system can reduce transmitting power by 25%. This characteristic is likely to be a potential advantage of the NLFM waveform, which can decrease the transmitting power requirement, particularly for radar satellites.

The high-resolution SAR processing algorithm customized for the NLFM chirp signal is crucial to the practical application of the NLFM waveform in SAR. A revised range migration algorithm (RMA) is developed to adapt it for focusing the NLFM SAR data in [6]. The Dechirping Compression Method for NLFM waveform was put forward in [7]. Saeedi et al. [8] explored the range-Doppler algorithm (RDA), omega-K (ω -K) algorithm, time domain correlation (TDC), and back-projection (BP) for the NLFM chirp signal. As a widely adopted high-resolution algorithm, the chirp scaling algorithm (CSA) was specifically designed for the interpolator utilized in range cell migration correction (RCMC) [9]. This algorithm is based on the scaling principle outlined by Papoulis [10], whereby a frequency modulation is applied to a chirp-encoded signal to realize signal shifting or scaling. Additionally, a modified imaging CSA incorporating two-step motion compensation (MoCo) was created to focus NLFM airborne SAR data [11].

A notable limitation of the NLFM waveform is its low Doppler tolerance. It is prone to the Doppler frequency shift caused by the relative motion between the radar system and targets. Notably, sonar systems are highly susceptible to such Doppler frequency shifts. Collins et al. [12] proposed hybrid-NLFM chirps, which merge NLFM with amplitude windowing; this allows Doppler sensitivity to be managed, albeit with a minor reduction in transmission efficiency. For spaceborne SAR, the Doppler frequency shift is also significant due to the high platform speed. Doerry, in a report for Sandia National Laboratories, concentrated on the impact of intrapulse Doppler in SAR signal processing and proposed a method to compensate for this effect [13]. Recently, a novel and effective compensation strategy has been introduced, in which a two-dimensional frequency domain model accounting for the Doppler shift effect is built to eliminate it [14].

The structure of this chapter is arranged as follows: Sect. 3.2 examines the signal-to-noise ratio (SNR) degradation associated with linear frequency-modulated (LFM) waveforms. Section 3.3 derives mathematical expressions for the output of matched filters, approaching the problem from both time-domain and frequency-domain perspectives separately. In Sect. 3.4, we introduce the revised range migration algorithm

(RMA) and chirp scaling algorithm (CSA) tailored to nonlinear frequency-modulated (NLFM) signals, which are applied to focus NLFM SAR data; additionally, a Doppler compensation scheme is detailed to counteract the impact of Doppler frequency shifts. Section 3.5 presents experimental results from airborne SAR tests, validating the performance of the proposed imaging algorithms and system error compensation methods. Lastly, Sect. 3.6 provides a summary of the chapter.

3.2 SNR Loss Problem

Normally, SAR system transmits a linear frequency modulation (LFM) waveform (e.g., TerraSAR-X [4, 5]) and pulse compression technology is employed to improve the processing gain. However, the instantaneous frequency of LFM waveform is a linear function of time and its power spectral density (PSD) is approximately rectangular. Therefore, with matched filtering (MF), the response function of LFM waveform is approximately a function with a peak-to-sidelobe ratio (PSLR) of −13.3 dB. This PSLR is usually considered to be too high, since it masks nearby weaker targets in the image. So additional method should be applied to reduce the sidelobes. Generally, weighted windowing [15] is applied to reduce the sidelobes. However, additional filtering makes the filtering process no longer precisely match to the signal, so the MF output would lose SNR, which greatly increase the power budget of spaceborne SAR system.

For a radar system, the SNR of the output at the time t_0 is

$$\mathrm{SNR}=\frac{|\int_{-\infty}^{+\infty}s(t_0-\tau)h(\tau)d\tau|^2}{N_0\int_{-\infty}^{+\infty}|h(\tau)|^2d\tau} \tag{3.1}$$

here, $s(t_0-\tau)$ corresponds to the radar pulse's echo signal, while t_0 denotes the pulse's time delay. $h(\tau)$ stands for the impulse response function, and N_0 refers to the noise power spectrum density.

Based on the matched filter (MF) principle, setting $h(\tau)=s^*(t_0-\tau)$maximizes the output signal-to-noise ratio (SNR), which is expressed as

$$\mathrm{SNR}_m=\frac{\int_{-\infty}^{+\infty}|s(t_0-\tau)|^2d\tau}{N_0} \tag{3.2}$$

In this equation, SNR_m represents the MF-induced output SNR.

When a weight window is introduced into the matched filter, the resulting SNR becomes

$$\mathrm{SNR}_w=\frac{|\int_{-\infty}^{+\infty}s(t_0-\tau)h_w(\tau)d\tau|^2}{N_0\int_{-\infty}^{+\infty}|h_w(\tau)|^2d\tau} \tag{3.3}$$

where $h_w(\tau)$ can be defined as

$$h_w(\tau) = W(\tau) \cdot h(\tau) = W(\tau) \cdot s^*(t_0 - \tau). \tag{3.4}$$

where $W(\tau)$ is the weight windowing. Accordingly, the output SNR after applying the weight window can be rephrased as

$$\mathrm{SNR}_w = \frac{|\int_{-T/2}^{+T/2} |s(t_0 - \tau)|^2 W(\tau) d\tau|^2}{N_0 \int_{-T/2}^{+T/2} |s(t_0 - \tau)|^2\, W(\tau)|^2 d\tau} \tag{3.5}$$

$$= \frac{|\int_{-T/2}^{+T/2} W(\tau) d\tau|^2}{N_0 \int_{-T/2}^{+T/2} |W(\tau)|^2 d\tau} \tag{3.6}$$

and the SNR without weight windowing is

$$\mathrm{SNR}_m = \frac{\int_{-T_r/2}^{+T_r/2} 1^2 d\tau}{N_0} = \frac{T_r}{N_0} \tag{3.7}$$

where T_r is the pulse duration. Thus, the SNR loss is defined as

$$\mathrm{SNR}_{loss} = 10 \log_{10} \frac{\mathrm{SNR}_w}{\mathrm{SNR}_m} = 10 \log_{10} \frac{|\int_{-T/2}^{+T/2} W(\tau) d\tau|^2}{T \int_{-T/2}^{+T/2} |W(\tau)|^2 d\tau}. \tag{3.8}$$

Yet, most existing studies [16–18], typically express this metric in a discrete format:

$$\mathrm{SNR}_{loss} = 10 \log_{10} \frac{(\sum_{n=1}^{N} w_n)^2}{N \sum_{n=1}^{N} w_n^2} \tag{3.9}$$

In this equation, w_n stands for the coefficients of the window function, while N refers to the total number of sampling points. Take the Hamming weighted filter as an instance: its corresponding SNR loss is 1.36 dB [17]. During satellite system design, this loss component must be accounted for, as it will require a higher transmission power.

3.3 Matched Filtering of NLFM Waveform

The purpose of this section is to derive expressions for matched filter output via time domain and frequency domain, respectively.

In detail, let the NLFM waveform be the transmitted signal $s(t)$. If the echo is received a time t_0 later, the target's echo can be expressed as

$$s_r(t) = \text{rect}\left(\frac{t - t_0}{T_r}\right) \exp\left(\text{j}\theta(t - t_0)\right)) \tag{3.10}$$

The time-domain matched filter is the time-reversed [19], complex conjugate of $s(t)$, with t_0 set to zero

$$h(t) = \text{rect}\left(\frac{t}{T_r}\right) \exp\left(-\text{j}\theta(-t)\right)) \tag{3.11}$$

The time-domain matched filter can be applied with a time domain convolution. It can also be applied with a frequency domain fast, giving the same results. In addition to being applied in the frequency domain, the matched filter can also be designed directly in the frequency domain, with little loss of accuracy [20]. The derivation starts with the same signal from a point target given in (3.10). Then, applying the POSP to (3.10), the signal spectrum is approximately given

$$S_r(f) \approx \text{rect}\left(\frac{f}{B_r}\right) \sqrt{W(f)} \exp\left(\text{j}\Phi(f)\right) \exp(-\text{j}2\pi f t_0) \tag{3.12}$$

where $W(f)$ is the weighting function, the additional linear phase term in the last exponential is due to the offset of the target t_0 from time 0, and $\Phi(f)$ denotes the phase of the NLFM spectrum, given by

$$\Phi(f) = \theta(t) - 2\pi f t, t = G(f) \tag{3.13}$$

In (3.13), $G(f)$ is the group time delay function. According to chain rule of derivation, it has

$$\begin{aligned} \frac{d\Phi(f)}{df} &= \frac{d\theta(t)}{dt}\frac{dt}{df} - 2\pi t - 2\pi f \frac{dt}{df} \\ &= -2\pi t = -2\pi G(f) \end{aligned} \tag{3.14}$$

Thus, the phase of the NLFM spectrum is proportional to the group time delay function. Then, the spectrum of the NLFM waveform can be recast as

$$S_r(f) \approx \text{rect}\left(\frac{f}{B_r}\right) \sqrt{W(f)} \exp\left(-\text{j}2\pi \int_{-\infty}^{f} G(f)\right) \exp(-\text{j}2\pi f t_0) \tag{3.15}$$

The matched filter is designed to cancel the integral phase term in (3.15), which is done by setting

$$H(f) = \text{rect}\left(\frac{f}{B_r}\right) \sqrt{W(f)} \exp\left(\text{j}2\pi \int_{-\infty}^{f} G(f)\right) \tag{3.16}$$

As an example, for Taylor weighting function, it has

$$W(f) = 1 + \sum_{m=1}^{\bar{n}-1} F_m \cos\left(\frac{2\pi m f}{B}\right), \quad |f| \leq B_r/2$$

$$G(f) = T\left(\frac{f}{B} + \frac{1}{2\pi}\sum_{m=1}^{\bar{n}-1} \frac{F_m}{m} \sin\left(\frac{2\pi m f}{B}\right)\right), \quad |f| \leq B_r/2 \tag{3.17}$$

Note that the integral phase term in (3.16) is independent of the target position t_0, which allows the cancellation. The spectrum of the signal, after the matched filter is applied, is given by

$$S_{out}(f) = S_r(f)H(f) = \text{rect}\left(\frac{f}{B_r}\right) W(f) \exp(-j2\pi f t_0) \tag{3.18}$$

The matched filter (3.11) can be applied in the time domain, using a linear convolution. However, as the matched filter is usually very long in SAR applications, the matched filter is normally applied in the frequency domain. There are two options to generate the frequency domain matched filter for NLFM waveform.

Option 1: general frequency-domain matched filter [11, 21]. This matched filter is generated by taking the complex conjugate of the discrete Fourier transform in the discrete time domain of the NLFM waveform, given by

$$H(f) = \{\mathbf{FT}(s(t))\}^* \tag{3.19}$$

where $\mathbf{FT}(\cdot)$ stands for the Fourier transform, while $(\cdot)^*$ indicates the complex conjugation operation.

Option 2: Analytical frequency-domain matched filter [20]. Generate the matched filter directly in the frequency domain by (3.16).

Figure 3.1 compares the pulse compression results of the analytical MF and the conventional MF, under the following parameters: signal bandwidth $B = 500\,\text{MHz}$, pulse duration $T = 10\,\mu\text{s}$, and a Taylor window with $\bar{n} = 5$, $sll = -35$. It can be seen that the two pulse compression outputs are highly similar. Additionally, the SNR of the pulse compression output was calculated via 1000 Monte Carlo simulation experiments, where the input SNR was set to 20 dB. Furthermore, the performance metrics of the pulse compression outputs are compiled in Table 3.1. In Table 3.1, it is shown that the analytical MF's metrics-including IRW, PSLR, ISLR, and output SNR-are all very close to those of the conventional MF. Thus, the analytical spectrum of the NLFM waveform is well-suited to serve as the matched filter for the corresponding NLFM signal.

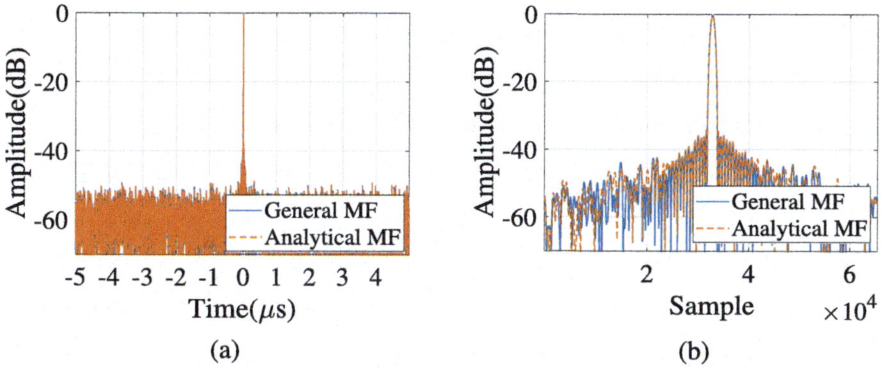

Fig. 3.1 MF outputs by the general and analytical MFs, respectively. **a** The MF output. **b** The corresponding interpolated plot

Table 3.1 Performance comparison between the general and analytical MFs

MF	IRW (sample)	PSLR (dB)	ISLR (dB)	SNR (dB)
General	1.185	−35.01	−20.67	58.83
Analytical	1.188	−34.64	−20.55	58.82

3.4 Imaging Algorithms and Doppler Compensation for NLFM-SAR System

In this section, firstly, the modified RMA and CSA for the NLFM signal are presented for focusing NLFM SAR data. Then the Doppler compensation approach is described to eliminate the Doppler shift effect.

3.4.1 Modified Range Migration Algorithm

For airborne SAR experiment, the accurate motion compensation [22–24] must be considered in the SAR processing. In this section, a modified extended RMA is developed to process NLFM chirp echoes for SAR imaging.

Taking the NLFM waveform as the transmitted pulse, and the baseband radar signal mixed up to the desired operating frequency before transmission. The transmitted NLFM signal can be expressed as

$$s_{pul}(t) = A_0 \text{rect}\left(\frac{t}{T_r}\right) \exp\left(j\left(2\pi f_c t + \theta(t) + \phi_0\right)\right) \tag{3.20}$$

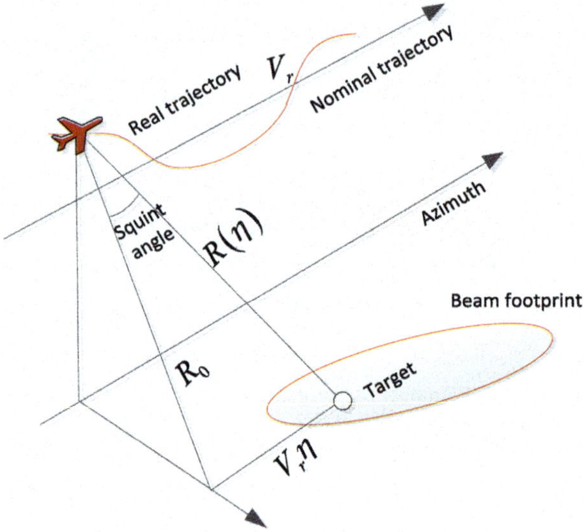

Fig. 3.2 Radar data acquisition geometry

where A_0 is a complex constant indicating the amplitude of the signal, f_c denotes the carrier frequency, and ϕ_0 is the starting phase, which can usually be neglected.

Figure 3.2 presents a basic geometric model describing the radar's position and its beam coverage area on the Earth's surface. Take a point target located R_0, aaway from the radar, characterized by a parameter A_1, which represents the backscatter coefficient. The signal received after the point target's reflection and subsequent down-conversion is

$$\begin{aligned} s(t,\eta) = A_1 \text{rect}\left(\frac{t-2R(\eta)/c}{T_r}\right) w_a(\eta-\eta_c) \\ \times \exp\left(-j4\pi f_c R(\eta)/c\right) \exp\left(j\theta(t-2R(\eta)/c)\right) \end{aligned} \tag{3.21}$$

where A_1 is a complex constant indicating the amplitude of the signal, η represents azimuth time, and η_0 is the beam center offset time. $R(\eta)$ represents the instantaneous slant range, $w_a(\eta)$ represents the azimuth envelope, and c is the speed of light. Considering a straight flight path, the slant range of a point target can be written as $R(\eta) = \sqrt{R_0^2 + V_r^2\eta^2}$, where R_0 is the slant range of the closest approach and V_r is the speed of the platform.

Performing a two-dimensional Fourier transform, the frequency domain of (3.21) can be expressed as

$$S(f, f_\eta) = A_2\sqrt{W(f)}W_a(f_\eta - f_{\eta_c})\exp(\mathrm{j}\theta_{2df}(f, f_\eta)) \tag{3.22}$$

The factor A_2 indicates the amplitude of the spectrum. $W_r(f)$ and $W_a(f_\eta)$ represent the frequency envelopes of the range and azimuth directions, respectively, and f_{η_c} is the Doppler centroid frequency. The phase term $\theta_{2df}(f, f_\eta)$ can be written as

$$\theta_{2df}(f, f_\eta) = -\frac{4\pi R_0}{c}\sqrt{(f_c + f)^2 - \frac{c^2 f_\eta^2}{4V_r^2}} + \Phi(f) \tag{3.23}$$

Then perform range compressing with frequency domain matched filter $H(f)$. After range compressing, $\Phi(f)$ is eliminated and the 2-D frequency spectrum of the processed signal is given by

$$S_1(f, f_\eta) = A_2 W(f) W_a(f_\eta - f_{\eta_c}) \exp\left(-\mathrm{j}\frac{4\pi R_0}{c}\sqrt{(f_c + f)^2 - \frac{c^2 f_\eta^2}{4V_r^2}}\right) \tag{3.24}$$

For conventional RMA, the next step is Stolt mapping, which performs the Range Cell Migration Correction (RCMC), secondary range compression (SRC), and azimuth compression. The range frequency is modified by replacing the square root factor in (3.24) with the shifted and scaled range frequency variable, which can be expressed as

$$f_1 = \sqrt{(f_c + f)^2 - \frac{c^2 f_\eta^2}{4V_r^2}} - f_c \tag{3.25}$$

In the case of airborne SAR, the implementation of MoCo commonly employs a two-step approach [22, 23] including range independent and range dependent components, respectively. The first step can be performed directly before or after range compression. The second step has to be applied after range compression and correction of Range Cell Migration (RCM). However, the conventional RMA is limited because of its inability to focus images with integrated and high-precision range dependent motion compensation. To overcome this, the modified Stolt mapping was first proposed in [25]. This modified Stolt mapping method separates RCMC and SRC from azimuth focusing, and the expression of the mapping can be described as

$$f_2 = \sqrt{(f_c + f)^2 - \frac{c^2 f_\eta^2}{4V_r^2}} - \sqrt{f_c^2 - \frac{c^2 f_\eta^2}{4V_r^2}} \tag{3.26}$$

where f_2 is the new range frequency variable. Following the adjusted Stolt mapping process, the data needs to be converted into the 2-D time domain-this enables the implementation of range-dependent motion compensation. Ultimately, azimuth compression can be carried out within the wide Doppler (RD) domain.

The block diagram of the proposed imaging algorithm tailored for NLFM waveform SAR is outlined in Fig. 3.3.

In order to validate the imaging algorithm for the NLFM waveform and demonstrate the characteristic of the NLFM chirp, the results of simulated raw data are

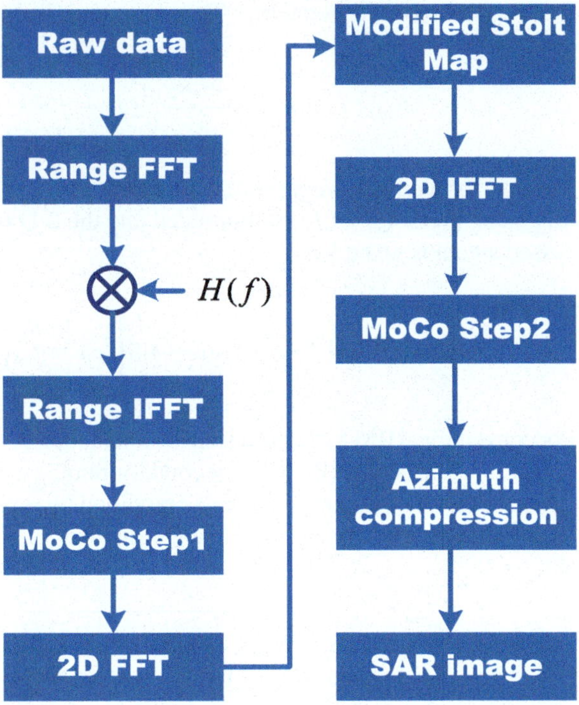

Fig. 3.3 Block diagram of modified imaging RMA for NLFM waveform

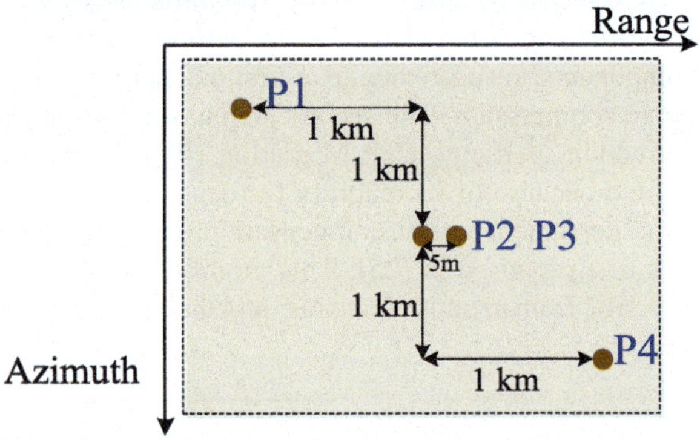

Fig. 3.4 The designed imaging scene

shown in this section. In the simulation, the NLFM waveform pulses are constructed to achieve a Taylor weighting function with a PSLR of −35 dB.

Based on the strip-map SAR imaging geometry, raw data corresponding to the NLFM waveform is simulated using the system parameters detailed in Table 3.1. The designed imaging scene is shown in Fig. 3.4 with four point targets. P2 and P3 are placed with short separation in range for analyzing the possible interference. It should be emphasized that the azimuth beam pattern is neglected in this simulation.

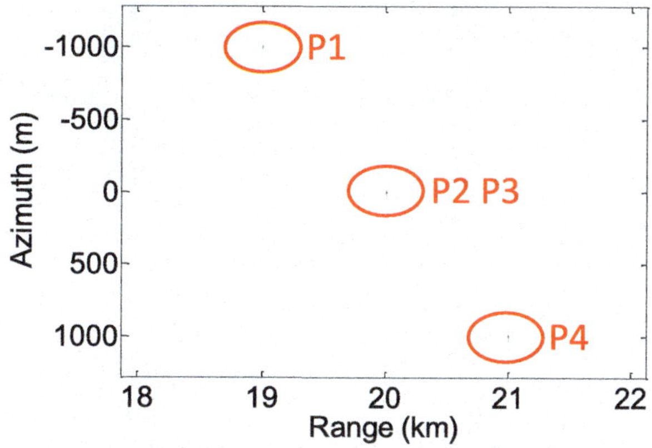

Fig. 3.5 The imaging result of the simulated data

The improved RMA was utilized to focus the simulated raw data, with the resulting outcome presented in Fig. 3.5. From Fig. 3.5, we can observe that the proposed imaging method effectively focuses all point targets.

Interpolated graphs of the point targets P1, P2, P3, and P4 are displayed in Fig. 3.6, and the corresponding performance assessments are listed in Table 3.2. Table 3.2 reveals that the range peak sidelobe ratios (PSLRs) of all point targets are close to 35 dB, and their range resolutions reach 1.44 m. For targets P2 and P3, mutual interference is minimal, and their PSLRs also approach −35 dB. To provide a comparison, an additional simulation using a linear frequency-modulated (LFM) waveform was conducted with the same system parameters listed in Table I. The range profiles from this LFM-based simulation are plotted as dotted lines in Fig. 3.6d, e and f. Notably, the measured range resolution for the LFM waveform is 1.01 m. From Fig. 3.6, it is evident that the sidelobes of the NLFM waveform are significantly lower than those of the LFM waveform. Thus, the proposed NLFM waveform and its corresponding imaging method are validated.

3.4.2 Modified Chirp Scaling Algorithm

This section develops the modified CSA for NLFM SAR data with motion compensation (MoCo). Firstly, by performing a range Fourier transform, (3.21) can be expressed as

$$\begin{aligned} s_r(f,\eta) = & A_3\sqrt{W(f)}w_a(\eta-\eta_c) \\ & \times \exp\left(-\mathrm{j}4\pi(f_c+f)R(\eta)/c\right)\exp\left(j\Phi(f)\right) \end{aligned} \tag{3.27}$$

where factor A_3 indicates the amplitude of the range spectrum. Range compression should be implemented first, the range frequency domain matched filter can be generated by taking the complex conjugate of discrete Fourier transform in the discrete

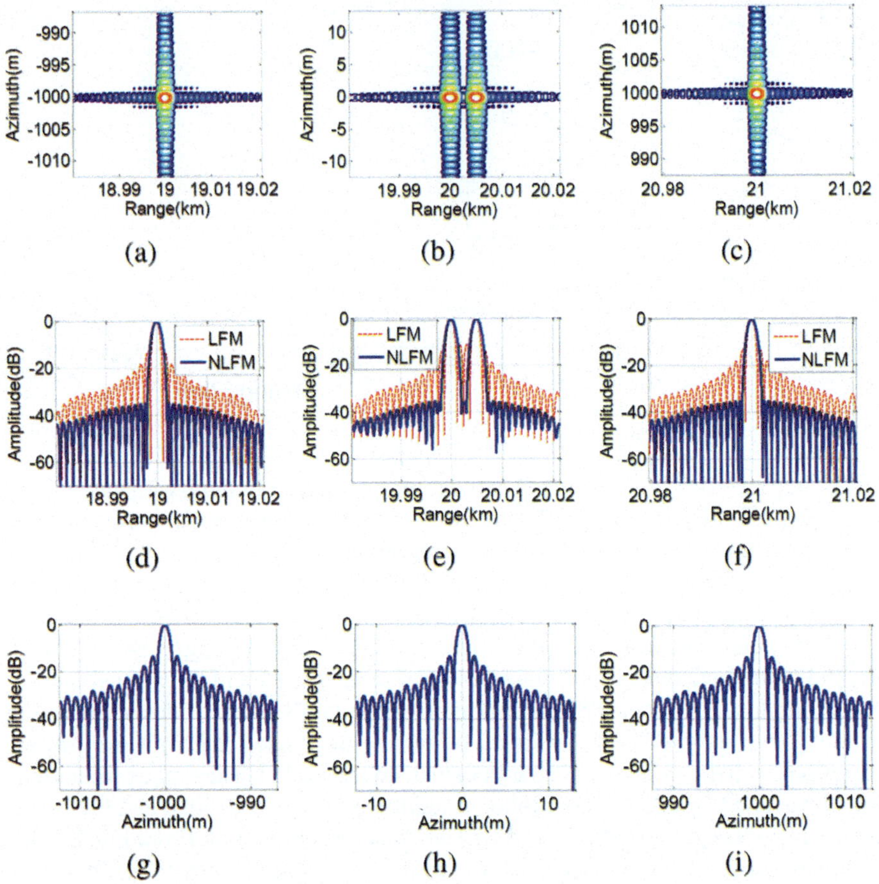

Fig. 3.6 The interpolated plots of the point targets P1, P2 and P3, P4. **a** The contour plot of P1; **b** the contour plot of P2 and P3; **c** the contour plot of P4; **d** the range profile of P1; **e** the range profile of P2 and P3; **f** the range profile of P4; **g** the azimuth profile of P1; **h** the azimuth profile of P2; **i** the azimuth profile of P4

time domain of the NLFM waveform. After range compression, $\Phi(f)$ is eliminated. In order to let the data have the range chirp encoding for the CSA, the data after range compression should be multiplied by the frequency spectrum of the LFM waveform as follows:

$$H_c(f) = \exp\left(-j\pi\frac{f^2}{K_r}\right) \tag{3.28}$$

where $K_r = B_r/T_r$ is the linear FM rate of the LFM waveform and (3.27) becomes

Table 3.2 Experiment system parameters

Parameters	Value
Platform velocity	125 m/s
Carrier frequency	9.6 GHz
Bandwidth	100 MHz
Pulse duration	10 μs
Pulse repetition frequency(PRF)	2999 Hz
DACs	1800 MHz
ADCs	1400 MHz

$$
\begin{aligned}
s_r(f, \eta) = & A_3 W(f) w_a(\eta - \eta_c) \\
& \times \exp\left(-\mathrm{j}4\pi (f_c + f) R(\eta)/c\right) \exp\left(-j\pi \frac{f^2}{K_r}\right).
\end{aligned} \tag{3.29}
$$

Performing a range inverse Fourier transform, (3.29) can be expressed as

$$
\begin{aligned}
s_r(t) = & A w(t - 2R(\eta)/c) w_a(\eta - \eta_c) \\
& \times \exp\left(-\mathrm{j}4\pi f_c R(\eta)/c\right) \exp\left(j\pi K_r (t - 2R(\eta)/c)^2\right).
\end{aligned} \tag{3.30}
$$

where $w(t) = W(f = K_r t)$ is the range envelop.

Then traditional CSA enables imaging. For airborne SAR, motion errors are large (from atmospheric turbulence and aircraft traits), so precise motion compensation [22–24] is critical for SAR processing. For airborne SAR, MoCo is usually implemented via a two-step method [26, 27], including range-independent and range-dependent components. The first step works before or after range compression, according to the Inertial Measurement Unit (IMU) information. In the second step of the imaging method in this paper, Phase Gradient Autofocus (PGA) was applied after range compression and RCMC. The block diagram of the modified CS imaging algorithm for the NLFM waveform SAR is summarized in Fig. 3.7.

Using strip-map SAR geometry, the NLFM waveform raw data are simulated with the system parameters listed in Table 3.2. The designed imaging scene is shown in Fig. 3.8 with nine point targets. It should be emphasized that the azimuth beam pattern is neglected in this simulation.

The proposed modified CSA is applied to focus the simulated raw data, and the corresponding result is shown in Fig. 3.9. Figure 3.9 shows that the point targets are well focused by the proposed imaging approach. The interpolated plots of point targets P1 and P5 are shown in Fig. 3.10, and the corresponding performance assessments are listed in Table 3.3. As shown in Table 3.3, the simulation results are quite close to the theoretical values, which validates the proposed NLFM waveform and imaging approach.

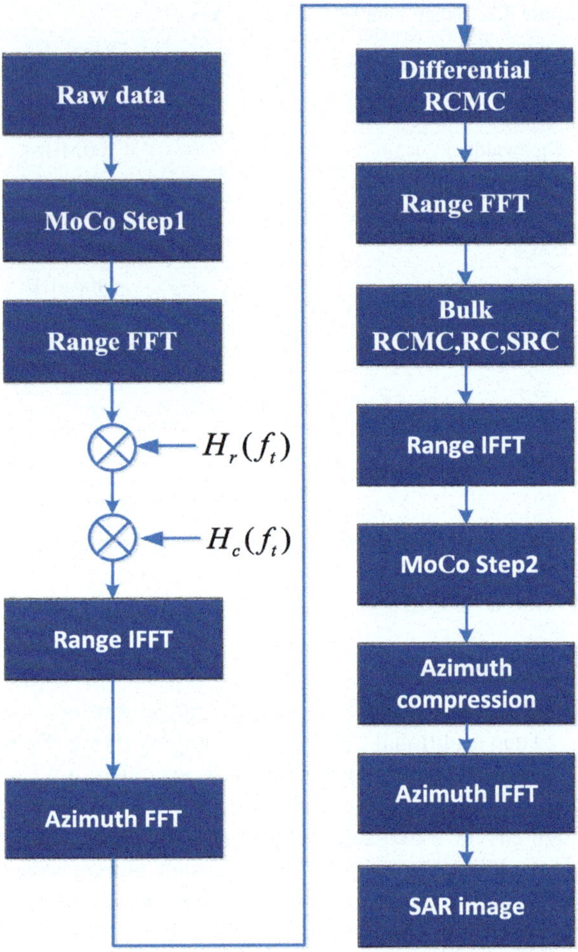

Fig. 3.7 Block diagram of the modified CSA for the NLFM waveform

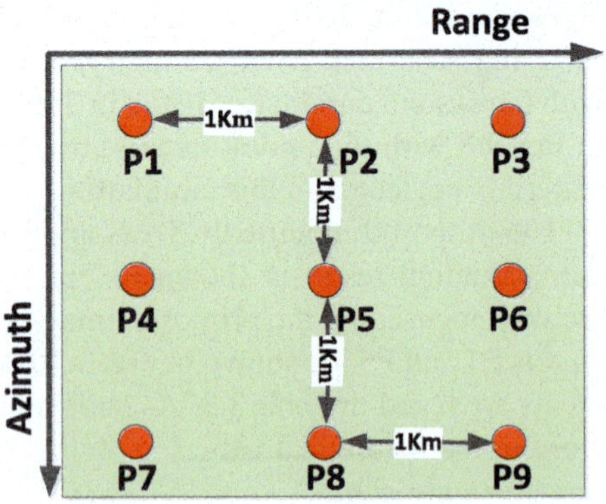

Fig. 3.8 Designed imaging scene

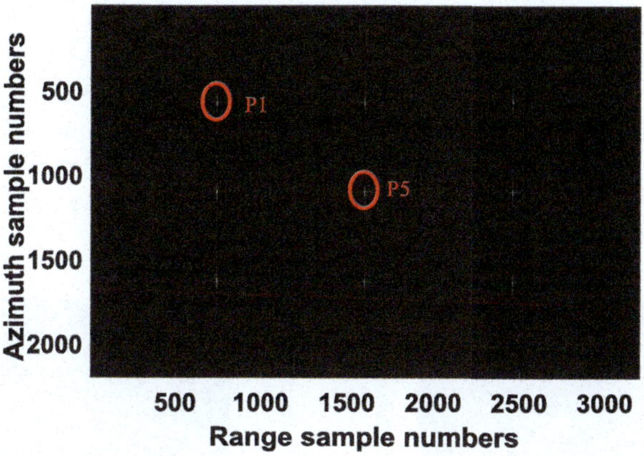

Fig. 3.9 Imaging result of the simulated data

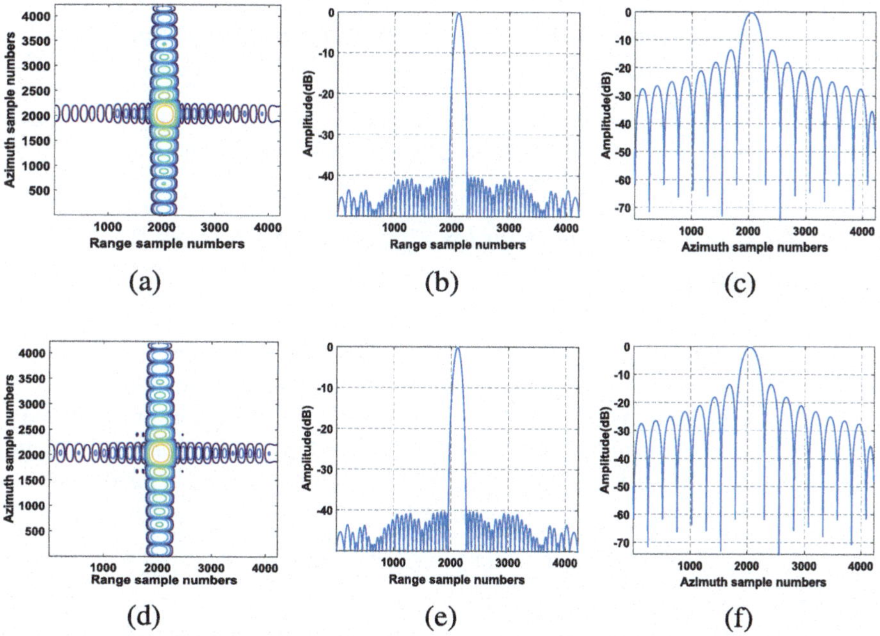

Fig. 3.10 Interpolated plots of point targets P1 and P5. **a** Contour plot of P1. **b** Range profile of P1. **c** Azimuth profile of P1. **d** Contour plot of P5. **e** Range profile of P5. **f** Azimuth profile of P5

Table 3.3 Imaging performance of point targets P1 and P5

Targets	Range			Azimuth		
	IRW (sample)	PSLR (dB)	ISLR (dB)	IRW (sample)	PSLR (dB)	ISLR (dB)
P1	1.17	−40.1	−28.3	0.89	−13.3	-9.5
P5	1.17	−40.1	−28.4	0.89	−13.3	-9.5

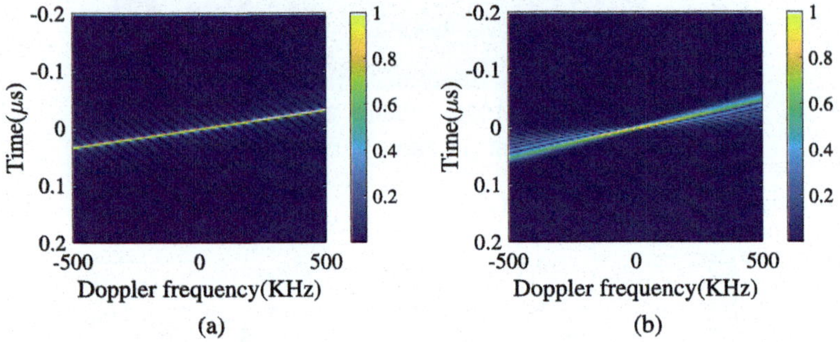

Fig. 3.11 Delay-Doppler ambiguity functions for waveforms (bandwidth: $B = 300$ MHz and pulse duration: $T = 20$ μs. **a** LFM waveform. **b** Taylor-harmonic NLFM waveform

3.4.3 Compensation for Doppler Effect ON NLFM SAR Imaging

In this section, the effect of Doppler shift on NLFM SAR imaging is assessed and one effective approach is proposed to compensate the effects on imaging.

With SAR the target is presumed stationary, however relative motion still exists due to the motion of the radar. Then, an intra-pulse Doppler shift is induced and may not be negligible. Specifically, the delay-Doppler ambiguity function is defined as

$$AF(t, f_d) = \left| \int_{-\infty}^{\infty} s(\tau)s^*(\tau - t)\exp(j2\pi f_d \tau)d\tau \right| \tag{3.31}$$

As an example, Fig. 3.11 illustrates the delay-Doppler ambiguity functions for the LFM waveform and NLFM waveform. Compared with the LFM waveform, obviously, the intra-pulse Doppler shift for NLFM waveform induces higher SNR loss relative the peak. Therefore, the effect of Doppler shift on NLFM SAR imaging cannot be neglected.

With the intra-pulse Doppler shift, the actual received SAR data is formulated as [14]

$$s(t,\eta) = w_r(t)(t)w_a(\eta)\exp\left(-\mathrm{j}4\pi\frac{R'(\eta)}{\lambda}\left(t-\frac{R(\eta)}{c}\right)\right)$$
$$\times\exp\left(-\mathrm{j}4\pi\frac{R(\eta)}{\lambda}\right)\exp\left(\mathrm{j}\theta\left(t-\frac{2R(\eta)}{c}\right)\right) \tag{3.32}$$

Accordingly, the range Fourier transform can be derived as

$$S_0(f,\eta) = \sqrt{W(f)}w_a(\eta)\exp\left(\mathrm{j}\Phi\left(f+\frac{2R'(\eta)}{\lambda}\right)\right)$$
$$\times\exp\left(-\mathrm{j}4\pi\frac{(f_c+f)R(\eta)}{c}\right) \tag{3.33}$$

The azimuth Fourier transform is given by

$$S_{2\mathrm{df}}(f,f_\eta) = \int_{-\infty}^{\infty} S_0(f,\eta)\exp(-\mathrm{j}2\pi f_\eta\eta)d\eta \tag{3.34}$$

Using the phase of $S_0(f,\eta)$ from (3.33), the phase of the Fourier integral is

$$\theta_d(\eta) = -4\pi\frac{(f_c+f)R(\eta)}{c} - \Phi\left(f+\frac{2R'(\eta)}{\lambda}\right) - 2\pi f_\eta\eta \tag{3.35}$$

Then the derivative of $\theta_d(\eta)$ with respect to η can be obtained

$$\frac{d\theta_d(\eta)}{d\eta} = -4\pi\frac{(f_c+f)R'(\eta)}{c} - 2\pi f_\eta$$
$$-4\pi G\left(f+\frac{2R'(\eta)}{\lambda}\right)\frac{R''(\eta)}{\lambda} \tag{3.36}$$

in which the first and second derivatives of $R(\eta)$ are expressed as

$$R'(\eta) = \frac{V_r^2\eta}{R(\eta)},\ R''(\eta) = \frac{V_r^2R_0^2}{R^3(\eta)} \leq \frac{V_r^2}{R(\eta)} \tag{3.37}$$

Further, it should be noted that, in general, since $G(f)$ and η are on the order of μs and s, respectively, it has

$$\left|G\left(f_t+\frac{2R'(\eta)}{\lambda}\right)\frac{R''(\eta)}{\lambda}\right| \ll \left|\frac{(f_c+f_t)R'(\eta)}{c}\right| \tag{3.38}$$

Thus, the last term of (3.36) can be neglected, and the derivative of $\theta_d(\eta)$ is approximately as

$$\frac{d\theta_d(\eta)}{d\eta} \approx -4\pi \frac{(f_c + f)R'(\eta)}{c} - 2\pi f_\eta \tag{3.39}$$

which is zero when

$$f_\eta = -2\frac{(f_c + f)R'(\eta)}{c} \tag{3.40}$$

Therefore, the phase of 2-D spectrum $S_{2\text{df}}(f_t, f_\eta)$ can be derived as

$$\theta_{2df}(f, f_\eta) = -\frac{4\pi R_0}{c}\sqrt{(f_c + f)^2 - \frac{c^2 f_\eta^2}{4V_r^2}} + \Phi\left(f - \frac{f_c f_\eta}{f_c + f}\right) \tag{3.41}$$

It is obvious that the 2-D spectrum suffers from a Doppler shift of $f_c f_\eta/(f_c + f)$. By expanding the Doppler shift in a power series of f, keep terms up to f, the Doppler shift becomes

$$\frac{f_c f_\eta}{f_c + f} \approx f_\eta - \frac{f_\eta}{f_c} f \tag{3.42}$$

Noted that $|f_\eta/f_c| \le 2V_r/c \ll 1$, the phase of 2-D spectrum $S_{2\text{df}}(f_t, f_\eta)$ can be reformulated as

$$\theta_{2df}(f, f_\eta) = -\frac{4\pi R_0}{c}\sqrt{(f_c + f)^2 - \frac{c^2 f_\eta^2}{4V_r^2}} + \Phi\left(f - f_\eta\right) \tag{3.43}$$

In order to eliminate the Doppler shift component, the last phase term of (3.41) can be removed using a compensation function $H_c(f, f_\eta)$, which can be obtained as

$$H_{com}(f, f_\eta) = \exp\left(-\mathrm{j}\Phi\left(f - f_\eta\right)\right) = \mathbf{FT}(s(t)\exp(\mathrm{j}2\pi f_\eta t)) \tag{3.44}$$

After removing the first exponential term in (3.43) using $H_{com}(f, f_\eta)$, the focusing algorithms introduced in previous section, such as modified CSA or RMA, can be used for imaging.

For a SAR system working with a squint angle, the Doppler shift magnitude exceeds that observed when the system operates in a zero-squint configuration. To reflect this characteristic, we use a squint spaceborne SAR in our simulation. Relevant system parameters are detailed in Table 3.4.

The 2-D frequency domain compensation function is applied to compensate the Doppler shift effect for SAR imaging. The focused imaging result with point targets P1, P2 and P3 for NLFM SAR with 15° beam squint angle is shown in Fig. 3.12. Figure 3.13 shows the contour results of point targets. From Fig. 3.13a, b and c, it can be seen that the imaging results are evidently distorted due to the high Doppler frequency shift. After compensating the Doppler frequency shift component using the compensation function, the point targets can be focused well, as shown in Fig. 3.13d,

Table 3.4 Simulation parameters

Parameter	Value
Height	500 km
Platform velocity	7600 m/s
Carrier frequency	5.3 GHz
Length of azimuth antenna	3 m
Doppler bandwidth	4489 Hz
Beam squint angle	15°
PRF	6080 Hz
Pulse width	10 μs
Bandwidth	100 MHz
Sampling rate	120 MHz

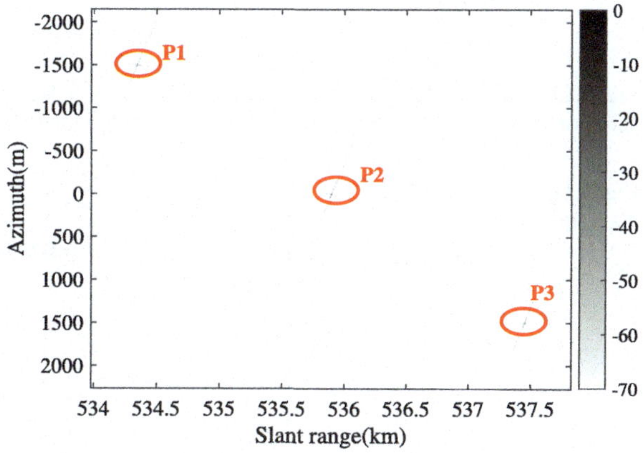

Fig. 3.12 Imaging result of the simulated data for NLFM SAR

e and f. Therefore, the deteriorated impacts of Doppler shift have been effectively compensated and the validity of the frequency domain compensation approach is demonstrated.

3.5 Real SAR System and Airborne Experiments

3.5.1 Real SAR System for the NLFM Waveform

In this section, we introduce the system structure and setup for the airborne experiment; the radar system in use was developed by the Institute of Electronics, Chinese Academy of Sciences. Next, we present the ground experiment and a system error compensation approach to correct amplitude-phase deviations. Lastly, we provide SNR evaluations of the ground experiment results.

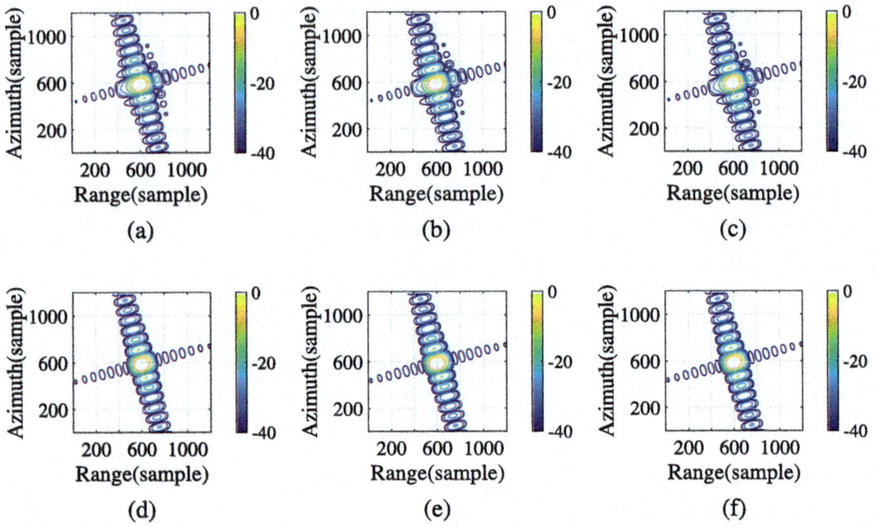

Fig. 3.13 Contour plots of point targets P1, P2 and P3. **a**, **b** and **c** Imaging results without Doppler shift compensation. **d**, **e** and **f** Imaging results with Doppler shift compensation

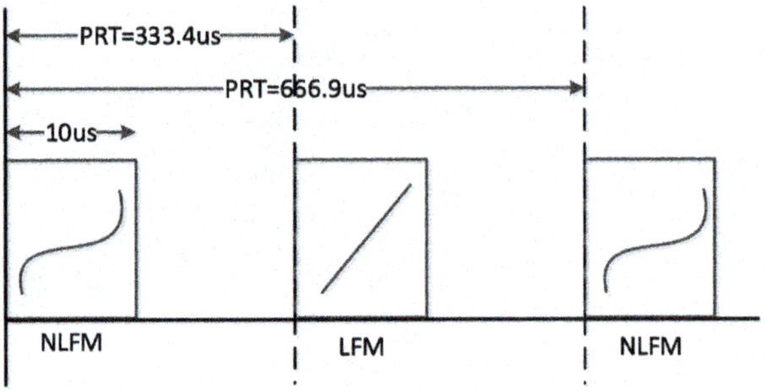

Fig. 3.14 Pulse sequence diagram

3.5.1.1 System Architecture and Configuration

This system is engineered to transmit LFM and NLFM waveforms alternately with identical parameters, and its pulse sequence diagram is shown in Fig. 3.14. This allows the antenna to illuminate the same area simultaneously, enabling SNR comparison between LFM and NLFM waveforms. As shown in Fig. 3.15, this radar system's structure mainly includes an antenna unit, electronic device unit, and data storage unit. The data storage unit uses a solid-state disk (SSD), while the antenna unit is a phased array antenna. The electronic device unit-made up of digital and RF subsystems (which integrate all electronics for signal generation/acquisition, control, and RF modulation/demodulation)-serves as the system's core.

In the digital subsystem, using the discrete values of the NLFM waveform, the baseband pulse is generated by FPGA. In the RF subsystem, the generated baseband

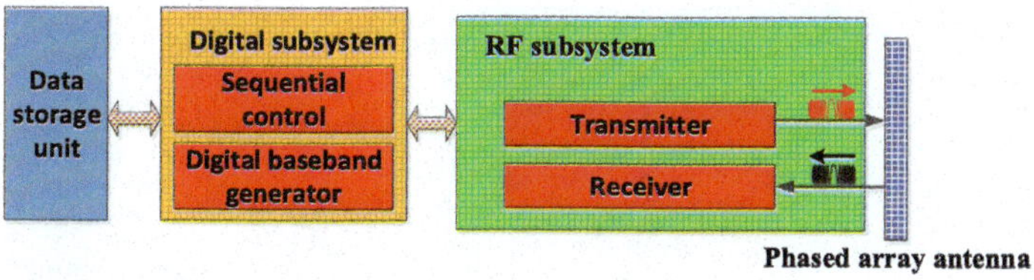

Fig. 3.15 Experiment system diagram

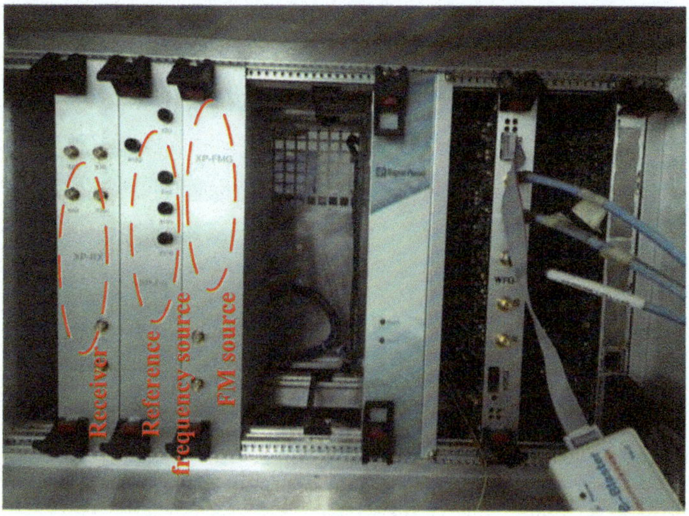

Fig. 3.16 Electronic device unit of the experiment system

NLFM signal is upconverted to an RF pulse, amplified by a high-power amplifier and then transmitted by the phased array antenna. The main parameters of the experiment system are listed in Table 3.2. Figure 3.16 shows the electronic device unit of the radar system.

A ground experiment plays a key role in initially validating the waveform's feasibility and extracting system errors for SAR data focusing. The designed experiment was conducted in a dark room, with the actual experimental setup shown in Fig. 3.17. A horn antenna was positioned about 12 m from the phased array antenna to receive the pulse; the received pulse then went through a delay line for a set time delay. After attenuation, the delayed signal was transmitted by the horn antenna to simulate an echo. This simulated echo was subsequently received by the phased array antenna and sampled by the system, with all timing in the digital subsystem controlled by the FPGA.

We conducted the ground experiment with the NLFM waveform without addressing system errors. The acquired baseband signal and its compressed outcome are displayed in Fig. 3.18. This outcome exhibits distortions, and the received signal's

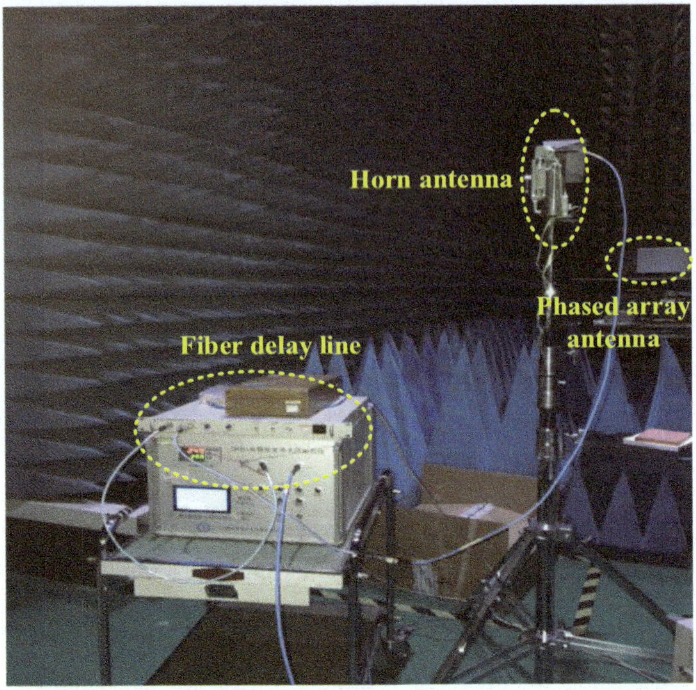

Fig. 3.17 Real ground experiment scenario

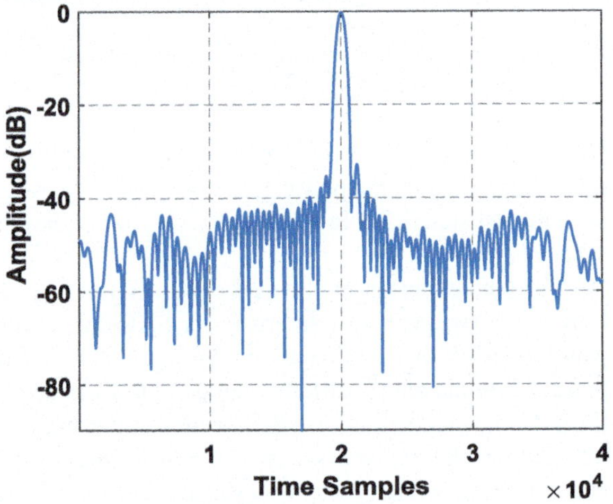

Fig. 3.18 Compressed result of the distorted pulse

performance falls short of theoretical expectations because real systems have inherent imperfections. The peak sidelobe dropped to −32 dB. Thus, compensating for system errors is necessary when focusing SAR data.

3.5.1.2 System Error Compensation Method

The signal model of the baseband distorted signal can be expressed as

$$s_d(t) = \mathrm{rect}(\frac{t}{T_r})a(t)\exp(j\theta(t) + j\phi(t)) \tag{3.45}$$

where $a(t)$ and $\phi(t)$ represent the the amplitude and phase distortion parts, respectively. By performing a Fourier transform, the frequency domain of the ideal baseband signal and distorted signal can be expressed as

$$S(f) = \mathbf{FT}[s(t)] = W(f)\exp(j\Theta(f)) \tag{3.46}$$

$$S_d(f) = \mathbf{FT}[s_d(t)] = A(f)\exp(j\Phi(f)) \tag{3.47}$$

where $W(f)$ and $\Theta(f)$ represent the envelope and the phase terms of $S(f)$, respectively, and $A(f)$ and $\Phi(f)$ represent the envelope and the phase terms of $S_d(f)$, respectively. To improve the SNR, statistical values should be measured using the average envelopes of multiple pulses while estimating the amplitude and the phase error as

$$\overline{A(f)} = \frac{1}{M}\sum_{i=1}^{M}|S_{d,i}(f)| \tag{3.48}$$

$$\overline{\Phi(f)} = \frac{1}{M}\sum_{i=1}^{M}\mathbf{phase}(S_{d,i}(f)) \tag{3.49}$$

where $S_{d,i}(f)$ represents the spectrum of the ith received pulse, M is the number of pulses and **phase**$(\cdot)$ represents phase unwrapping.

Then the system error compensation term can be obtained as

$$D(f) = \frac{W(f)}{\overline{A(f)}}\exp(j\Theta(f) - j\overline{\Phi(f)}). \tag{3.50}$$

This experiment is performed again using (3.50) by multiplying the collected signal in the frequency domain, and the compressed result is shown in Fig. 3.19. The compressed signal has about peak sidelobe of −39.2 dB. Therefore, the characteristics are close to the theoretical value of −40.2 dB and the recovery of the performance of the received signal is evident. Based on the experimental results, the error compensation method is effective for the test radar system.

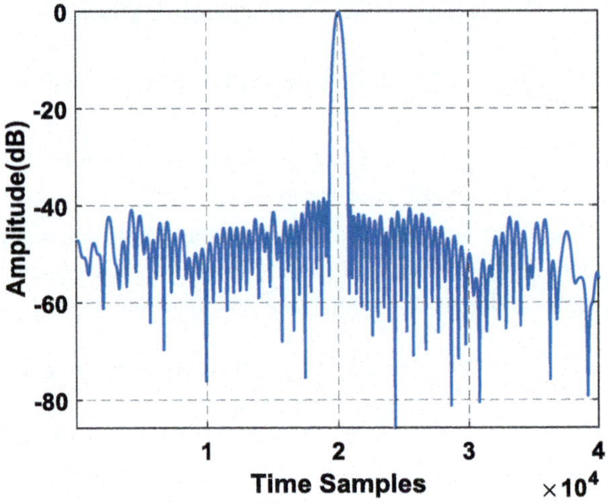

Fig. 3.19 Compression result of the received pulse in the experiment after compensating the system error

3.5.1.3 Evaluation of SNR

A test involving NLFM and LFM waveforms was conducted to showcase the merits of the NLFM waveform. In this test, the two waveforms share the same bandwidth and pulse duration, namely, $B = 100$ MHz and $T = 10\,\mu$s. To obtain the same sidelobes, the received LFM pulse is weighted by a -42 dB Taylor weighting function that has the same sidelobe height as that of the NLFM waveform at approximately -39 dB.

We processed ten pulses from each waveform, with the assessment results plotted in Fig. 3.20. Following range compression, the LFM pulses achieved an average SNR of about 70.12 dB, while the NLFM pulses reached an average of roughly 71.36 dB. Compared to the LFM waveform, the NLFM variant offers a 1.24 dB SNR gain-this initially confirms the NLFM waveform's advantages.

3.5.2 *Airborne SAR Experiment*

This experiment uses an X-band SAR system that alternates between transmitting an advanced NLFM waveform and an LFM waveform with the same parameters, is operated in strip-map mode. This SAR system is mounted on an airborne platform, a small Cessna plane, as shown in Fig. 3.21. The phased array antenna cover sits beneath the aircraft fuselage-this area is marked by the red rectangle in Fig. 3.21. The electronic equipment is housed inside the aircraft, and the airborne SAR system's parameters are detailed in Table 3.2.

As shown in Fig. 3.22, after collecting the echo signal, we first extract the LFM and NLFM signal data separately. Then, the system error compensation term (3.50) is applied to multiply the raw data in the range frequency domain to compensate

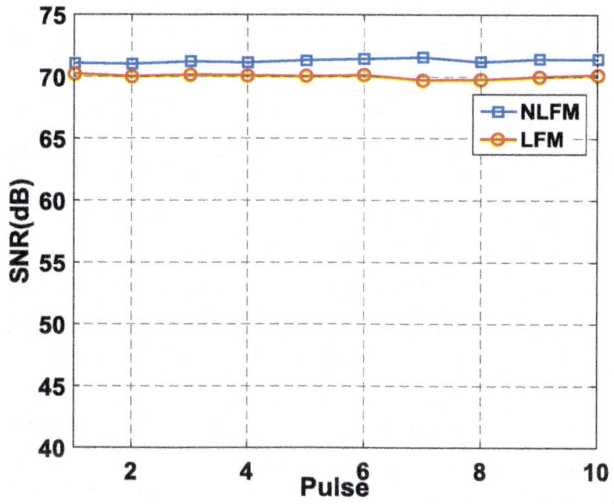

Fig. 3.20 Evaluated SNRs of the LFM and NLFM waveforms

Fig. 3.21 Employed airborne platform

the system error. Finally, the modified CSA with two-step MoCo is performed for SAR imaging. The focused image obtained from the experimental data is presented in Fig. 3.23.

Additionally, imaging outcomes for the LFM waveform, the LFM waveform paired with a −42 dB Taylor weighting function and the proposed NLFM waveform are shown in Fig. 3.24a–c. To analyze the results, the strong point targets marked in the rectangular region are shown in Fig. 3.24d–f. It is clearly shown that the point targets are well focused. The range profile of the point targets is shown in Fig. 3.25. Table 3.5 quantifies the IRW and PSLR of the range profile results. These results

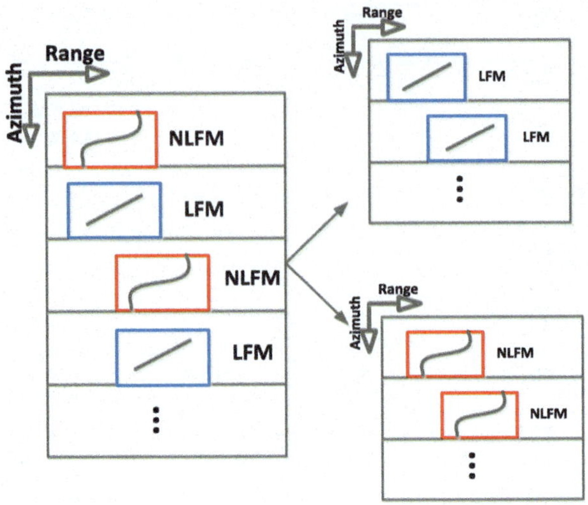

Fig. 3.22 SAR data extraction

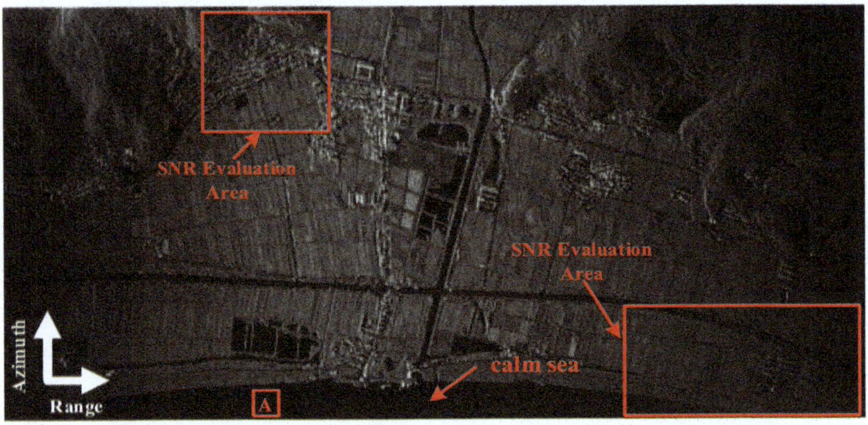

Fig. 3.23 SAR image (Ningbo, China) processed by CSA

shows that the proposed NLFM waveform can reach the same sidelobe as the LFM waveform with a −42 dB Taylor weighting function, while reducing the main lobe. In addition, compared with LFM waveform with a −35 dB Taylor weighting function, the proposed NLFM waveform can reduce the sidelobe by approximately 5 dB without broadening the main lobe. Actually, compared with the compressed results in the ground experiment, the imaging results have higher sidelobes because the system error is unstable. However, as shown in Fig. 3.26, after error compensation, the imaging results indicate that the sidelobes can reduce from −27 dB to −31 dB, which verifies the availability of the error compensation scheme. Furthermore, we only care about the difference between the initial waveform and the optimized sig-

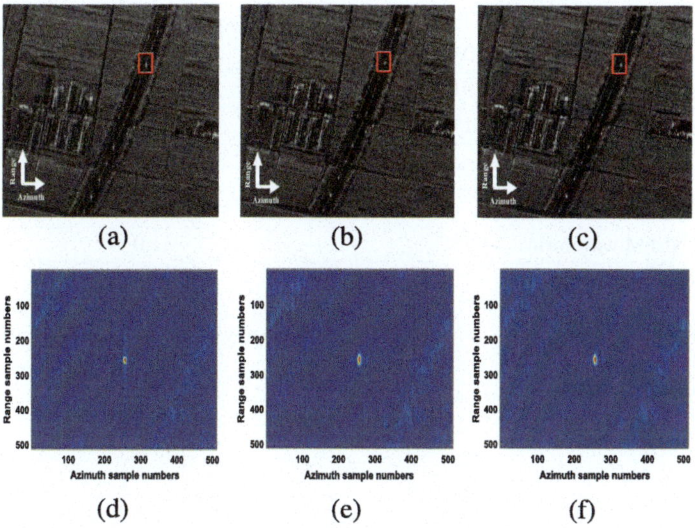

Fig. 3.24 Imaging results of different waveforms. **a** LFM waveform. **b** LFM waveform with a −42 dB Taylor weighting function. **c** Proposed NLFM waveform. **d–f** The strong scatter point targets marked

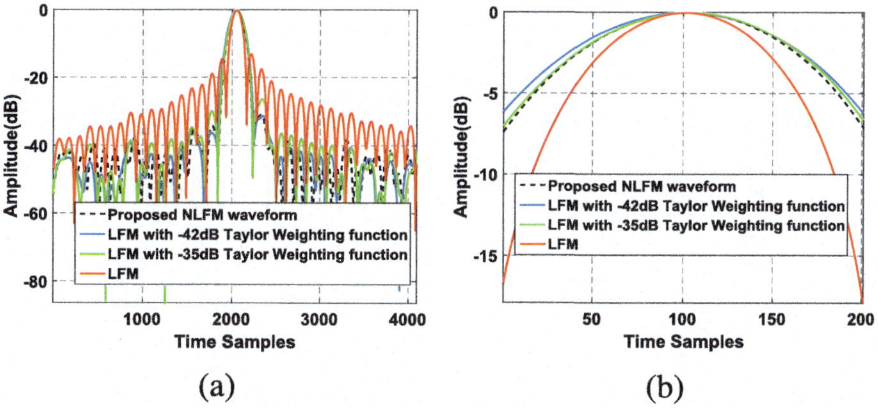

Fig. 3.25 Interpolated results for the strong scatter point targets marked in Fig. 3.24. **a** Range profile of the interpolated result. **b** Range profile of the interpolated result (close up)

nal. The experimental results are enough to verify the feasibility of the optimization framework.

The peak signal to noise ratio (PSNR) [28, 29] is widely used to evaluate the image SNR

$$\mathrm{SNR} = 10\log_{10}(\frac{S_p^2}{E[n(t,\eta)]}) \tag{3.51}$$

Table 3.5 Quantified performance for the range profile of the point targets

Frequency modulation	IRW (sample)	PSLR (dB)
LFM waveform	0.91	−12.9
LFM with −35 dB Taylor weighting function	1.19	−26.1
LFM with −42 dB Taylor weighting function	1.28	−31.1
Optimized NLFM waveform	1.18	−31.0

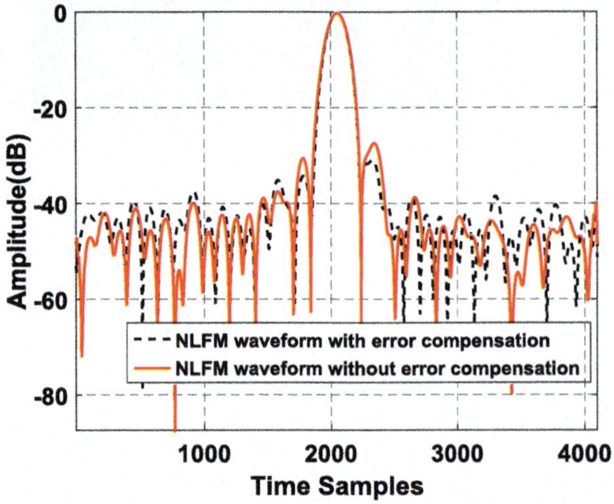

(a)

Fig. 3.26 Range profile of the interpolated result with error compensation versus without error compensation

where $E[n(t, \eta)]$ is the average power of the noise

$$E[n(t, \eta)] = \frac{1}{MN} \sum_{i=1}^{M} \sum_{j=1}^{N} N(i, j) \tag{3.52}$$

where $N(i, j)$ is the image noise. Area A in Fig. 3.23 corresponds to the calm sea in the received echo, and this region can be treated as image noise. So we utilize a 128×128 windowed region-labeled A-to compute the average noise power. S_p stands for the maximum pixel value within the image. Four distinct regions are selected to assess the image's SNR, as illustrated in Fig. 3.27, with the assessment outcomes presented in Table 3.6. Areas A, B and C are targets located on the highway. In addition, B and C are isolated targets, so their SNR values are nearly identical. Area D consists of dense scattering points, making it the region with the highest SNR. Yet the SNR enhancement observed in the image is roughly consistent across these regions, with an average gain of 1.29 dB-this aligns with the SNR assessment result

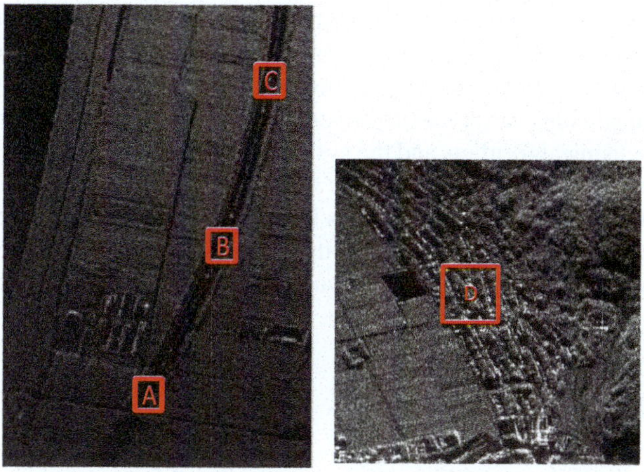

(a) (b)

Fig. 3.27 SNR evaluation areas. **a** SNR evaluation areas on the highway. **b** SNR evaluation area constituted by some dense scattering points

Table 3.6 SAR image SNR evaluation results (dB)

Area	SNR for NLFM	SNR for LFM	SNR improvement
A	72.72	71.4	1.32
B	67.42	66.13	1.29
C	66.68	65.4	1.28
D	75.45	74.18	1.27

from the ground experiment. This further confirms that the NLFM waveform offers the benefit of boosting SNR.

3.6 Summary

The NLFM waveform serves as a promising alternative to the conventional LFM waveform, as it enables sidelobe reduction without compromising signal-to-noise ratio. In this chapter, we first derive the time-domain and frequency-domain expressions of the matched filter tailored for the NLFM waveform. To effectively focus NLFM airborne SAR data, we further propose modified range migration algorithm and chirp scaling algorithm that integrate two-step motion compensation. To mitigate the impact of Doppler shift on NLFM SAR imaging, a dedicated Doppler compensation method is developed with simulation results confirming its validity. We then introduce a practical SAR system designed for NLFM waveform operation, alongside a ground experiment and a corresponding system error compensation strategy. Finally, an airborne SAR experiment is conducted where the LFM waveform and the optimized NLFM waveform are transmitted alternately. Both simulation and experimental imaging results verify the effectiveness of the proposed NLFM waveform the developed focusing algorithms and the system error compensation scheme. Notably, the NLFM waveform is found to enhance signal-to-noise ratio by 1.29 dB compared with the traditional LFM waveform. This improvement translates to more than 25% savings in output power. The findings of this paper thus demonstrate that the NLFM waveform holds substantial potential for application in future SAR systems.

References

1. Curlander JC, Mcdonough RN. Synthetic Aperture Radar: Systems and Signal Processing. New York, NY, USA; 1991.
2. Krieger G, Moreira A, Fiedler H, et al. TanDEM-X: A Satellite Formation for High-Resolution SAR Interferometry. IEEE Transactions on Geoscience and Remote Sensing. 2007;45(11):3317–3341.
3. Xu F, Song Q, Jin YQ. Polarimetric SAR Image Factorization. IEEE Transactions on Geoscience and Remote Sensing. 2017 Sept;55(9):5026–5041.
4. Meta A, Mittermayer J, Prats P, et al. TOPS Imaging With TerraSAR-X: Mode Design and Performance Analysis. IEEE Transactions on Geoscience and Remote Sensing. 2010;48(2):759–769.
5. Brusch S, Lehner S, Fritz T, et al. Ship Surveillance With TerraSAR-X. IEEE Transactions on Geoscience and Remote Sensing. 2011;49(3):1092–1103.
6. Wang W, Wang R, Zhang Z, et al. First Demonstration of Airborne SAR With Nonlinear FM Chirp Waveforms. IEEE Geoscience and Remote Sensing Letters. 2016;13(2):247–251.
7. Liu J, Zhang Y, Dong X. Dechirping Compression Method for Nonlinear Frequency Modulation Waveforms. IEEE Geoscience and Remote Sensing Letters. 2018;p. 1–5.
8. Saeedi J, Faez K. Synthetic aperture radar imaging using nonlinear frequency modulation signal. IEEE Transactions on Aerospace and Electronic Systems. 2016;52(1):99–110.
9. Runge H, Bamler R. A Novel High Precision SAR Focussing Algorithm Based On Chirp Scaling. In: Geoscience and Remote Sensing Symposium, 1992. IGARSS '92. International. vol. 1; 1992. p. 372–375.
10. Cumming I, Wong F, Raney K. A SAR Processing Algorithm With No Interpolation. In: Geoscience and Remote Sensing Symposium, 1992. IGARSS '92. International. vol. 1; 1992. p. 376–379.
11. Jin G, Deng Y, Wang R, et al. An Advanced Nonlinear Frequency Modulation Waveform for Radar Imaging With Low Sidelobe. IEEE Transactions on Geoscience and Remote Sensing. 2019;57(8):6155–6168.
12. Collins T, Atkins P. Nonlinear frequency modulation chirps for active sonar. IEE Proceedings - Radar, Sonar and Navigation. 1999 ;146(6):312–316.
13. Doerry AW. SAR processing with non-linear FM chirp waveforms. Sandia National Laboratories; 2006.
14. Wang W, Wang R, Deng Y, et al. Demonstration of NLFM Waveforms With Experiments and Doppler Shift Compensation for SAR Application. IEEE Geoscience and Remote Sensing Letters. 2016;13(12):1999–2003.
15. Shayesteh MG, Mottaghi-Kashtiban M. FIR filter design using a new window function. In: 2009 16th International Conference on Digital Signal Processing; 2009. p. 1–6.
16. Kurdzo JM, Cheong BL, Palmer RD, et al. Optimized NLFM pulse compression waveforms for high-sensitivity radar observations. In: 2014 International Radar Conference; 2014. p. 1–6.

17. Levanon N, Mozeson E. Radar signals. John Wiley & Sons; 2004.
18. Richards MA. Fundamentals of radar signal processing. Tata McGraw-Hill Education; 2005.
19. Cumming IG, Wong FH. Digital processing of synthetic aperture radar data: algorithms and implementation. Norwood, MA, USA: Artech House; 2005.
20. Zhang Y, Deng Y, Zhang Z, et al. Analytic NLFM Waveform Design with Harmonic Decomposition for Synthetic Aperture Radar. IEEE Geoscience and Remote Sensing Letters. 2022;19:1–5.
21. Wang W, Wang R, Zhang Z, et al. First Demonstration of Airborne SAR With Nonlinear FM Chirp Waveforms. IEEE Geoscience and Remote Sensing Letters. 2016;13(2):247–251.
22. Moreira A, Mittermayer J, Scheiber R. Extended chirp scaling algorithm for air- and spaceborne SAR data processing in stripmap and ScanSAR imaging modes. IEEE Transactions on Geoscience and Remote Sensing. 1996;34(5):1123–1136.
23. Moreira JR. A New Method Of Aircraft Motion Error Extraction From Radar Raw Data For Real Time Motion Compensation. IEEE Transactions on Geoscience and Remote Sensing. 1990;28(4):620–626.
24. Fornaro G, Franceschetti G, Perna S. Motion compensation errors: effects on the accuracy of airborne SAR images. IEEE Transactions on Aerospace and Electronic Systems. 2005;41(4):1338–1352.
25. Reigber A, Potsis A, Alivizatos E, et al. Extended wavenumber domain SAR focusing with integrated motion compensation. In: IGARSS 2003. 2003 IEEE International Geoscience and Remote Sensing Symposium. Proceedings(IEEE Cat. No.03CH37477). vol. 3; 2003. p. 1465–1467.
26. Moreira A, Mittermayer J, Scheiber R. Extended chirp scaling algorithm for air- and spaceborne SAR data processing in stripmap and ScanSAR imaging modes. IEEE Transactions on Geoscience and Remote Sensing. 1996 ;34(5):1123–1136.
27. Moreira JR. A New Method Of Aircraft Motion Error Extraction From Radar Raw Data For Real Time Motion Compensation. IEEE Transactions on Geoscience and Remote Sensing. 1990;28(4):620–626.
28. Gupta A, Tripathi A, Bhateja V. Despeckling of SAR Images via an Improved Anisotropic Diffusion Algorithm. In: Proceedings of the International Conference on Frontiers of Intelligent Computing: Theory and Applications (FICTA). Berlin, Heidelberg: Springer Berlin Heidelberg; 2013. p. 747–754.
29. Bhateja V, Tripathi A, Gupta A. An Improved Local Statistics Filter for Denoising of SAR Images. In: Recent Advances in Intelligent Informatics: Proceedings of the Second International Symposium on Intelligent Informatics (ISI'13), August 23-24 2013, Mysore, India. Cham: Springer International Publishing; 2014. p. 23–29.

Chapter 4
Quasi-Orthogonal Waveform Design for Range Ambiguity Suppression

4.1 Introduction

With the continued development of high-resolution radar satellites for global environmental observation, a variety of spaceborne SAR missions including TerraSAR-X, COSMO-SkyMed and Sentinel-1 [1–4] have been deployed and are widely applied in remote sensing. Despite these notable achievements, present-generation SAR sensors still encounter two fundamental constraints. The first is the inability of a single system to simultaneously deliver fine spatial resolution and broad swath coverage. The second stems from the intrinsic balance between range and azimuth ambiguities [5]. When the pulse repetition frequency (PRF) is set too high, echoes from far ranges may return during the transmission of the following pulse, leading to temporal overlap between desired and interfering signals. This overlap introduces extra energy, known as range ambiguity, which deteriorates SAR image quality. Existing techniques for mitigating range ambiguities can generally be grouped into two major classes.

- *More flexible SAR systems, e.g., multichannel SARs* [6–8] *and Multiple-Input Multiple-Output (MIMO) SARs* [9–12];
- *Advanced transmission schemes together with a corresponding signal processing method on receive* [13–17].

4.1.1 Background and Previous Works

A wide range of approaches has been developed to suppress range ambiguities in radar systems, particularly in Synthetic Aperture Radar (SAR). One representative method is the Azimuth Phase Coding (APC) technique proposed by Dall and Kusk [14]. By applying azimuth-dependent phase modulation to the transmitted signal, APC shifts

R. Wang and W. Wang, *Synthetic Aperture Radar (SAR) Waveform Design and Signal Processing*, https://doi.org/10.1007/978-981-95-7549-7_4

range-ambiguous components into the azimuth dimension, where they can be attenuated through filtering. However, this method requires a relatively high pulse repetition frequency (PRF) to maintain sufficient separation between the desired echo and the ambiguous energy, which considerably limits the swath that can be achieved. An alternative approach is to transmit different waveforms in consecutive Pulse Recurrence Intervals (PRIs), forming the basis of the Alternate Transmitting Mode (ATM) first introduced in [18]. This mode is well suited for conventional SAR systems with one transmit channel and one receive channel, because its ambiguity suppression capability relies on the low cross-correlation energy (CCE) between the alternating waveforms, which enables effective suppression without relying on PRF and therefore makes it attractive for spaceborne SAR applications. ATM can be implemented through frequency diversity or code diversity. In the frequency-diversity configuration, waveforms with different carrier frequencies are transmitted in different PRIs, and the reduced spectral overlap results in low residual range ambiguities, although the loss of coherence generates strong azimuth ambiguities [18]. In the code-diversity configuration, quasi-orthogonal waveforms with the same carrier frequency are used, and the range ambiguities are reduced after range compression. Consequently, the design of suitable quasi-orthogonal waveform sets has become a major topic in recent studies [19–21].

A technique that employs alternating up-chirp and down-chirp transmissions was investigated in [22, 23]. After range compression, however, the cross-correlation energy (CCE) between these two types of waveforms becomes either smeared or shifted to different positions in the range and azimuth dimensions [9, 24]. When the observed scene contains distributed targets, the mismatched ambiguity components accumulate across the image, which causes the overall ambiguity energy to remain at a high level. As a result, up-chirp and down-chirp signals cannot effectively mitigate range ambiguities because they exhibit essentially identical CCE characteristics [9, 22]. To address this problem, more sophisticated waveform designs have been proposed, such as Orthogonal Frequency Division Multiplexing (OFDM) chirp waveforms [19, 20]. Although these waveforms provide improved orthogonality, they introduce strong recurrent lobes, which make them unsuitable for wide-swath spaceborne SAR imaging and prevent effective suppression of ambiguity energy [9].

In summary, ensuring waveform orthogonality in SAR is more demanding than in communication systems, as the radar must sustain robust imaging performance over distributed targets. To suppress ambiguity energy effectively, the quasi-orthogonal waveforms applied in SAR should possess low cross-correlation energy (CCE). Theoretical results show that waveforms with flat spectra yield a constant CCE for any waveform pair [9], implying that non-flat spectral characteristics are required when seeking further CCE reduction. In addition, because SAR operates as an imaging sensor, the transmitted signals must maintain low Peak-to-Side-Lobe Ratios (PSLRs) and low Integrated Side-Lobe Ratios (ISLRs) to guarantee image quality. In this regard, the quasi-orthogonal Nonlinear Frequency Modulation (NLFM) waveforms designed for quad-pol SAR in [25], which simultaneously provide low PSLR and decreased CCE, offer a compelling candidate.

4.1.2 Main Contribution and Organization

This study offers refined insights into the ambiguity behavior of the Alternate Transmitting Mode (ATM) in SAR, establishes a detailed signal model that can assist system designers in optimizing ambiguity performance for spaceborne SAR, and examines the use of quasi-orthogonal NLFM waveforms in the ATM scheme to enhance range-ambiguity suppression. The primary contributions of this work are outlined as follows:

- Although [26] offered an initial discussion on the imaging effects introduced by waveform diversity, it did not provide a rigorous and complete signal model. The present work addresses this limitation by establishing a detailed signal model that enhances the understanding of this operating mode.
- Based on the established signal model, the azimuth ambiguity caused by waveform diversity is thoroughly analyzed. To mitigate the additional azimuth ambiguity, this paper proposes an azimuth compensation method that leverages *priori* knowledge of the transmitted waveforms.
- Previous studies [19–21] have examined the Range Ambiguity-to-Signal Ratio (RASR) in ATM, but they neglect the effect of distributed target scenes. To overcome this limitation, this paper presents more general and precise RASR formulations that consider waveform diversity.
- By employing quasi-orthogonal NLFM waveforms that feature reduced sidelobes and diminished CCE, this work provides an in-depth assessment of the ambiguity-mitigation performance of the ATM scheme.

The remainder of this paper is arranged as follows. Section 4.2 outlines the optimization framework adopted for designing orthogonal waveforms. Section 4.3 presents the ATM signal model and analyzes the azimuth ambiguities introduced by waveform diversity, along with the associated RASR expressions. Section 4.4 introduces a practical approach to mitigate the extra azimuth ambiguities induced by diverse waveforms. With reference to the LuTan-1 (LT-1) mission [27], Sect. 4.5 describes the integration of quasi-orthogonal NLFM signals into the ATM scheme and provides extensive simulation results that support the analytical findings. Section 4.6 offers concluding remarks.

Notation: Bold uppercase and lowercase letters represent matrices and column vectors, respectively. The operator $(\cdot)^*$ denotes the complex conjugate. $\mathscr{F}t\cdot$ $(\mathscr{F}^{-1}t\cdot)$ stands for the (inverse) Fourier transform with respect to fast time, while $\mathscr{F}\eta\cdot$ $(\mathscr{F}^{-1}\eta\cdot)$ refers to the (inverse) Fourier transform with respect to slow time. $x \pmod y$ indicates the remainder of x modulo y, $|\cdot|_2$ denotes the 2-norm, and $\odot$ represents the Hadamard product.

4.2 Quasi-Orthogonal Waveform Design

4.2.1 Signal Model

For waveform optimization, it is essential to adopt a clear parameterization of the instantaneous frequency. While B'ezier curves [28] offer a convenient means of describing the signal, their high smoothness may restrict the design flexibility. A symmetric profile based on piecewise linear (PWL) functions was used in [29]. Since symmetry is not a necessary property of the optimal frequency law, this study instead utilizes asymmetric PWL modeling. Accordingly, the instantaneous frequency is constructed with $2n + 2$ linear segments associated with $2n$ tunable control points, as depicted in Fig. 4.1.

The instantaneous frequency of the designed NLFM waveform is formulated as follows:

$$f_{+}(t) = \begin{cases} k_{10}t & 0 \le t < T_{11} \\ B_{11} + k_{11}(t - T_{11}) & T_{11} \le t < T_{12} \\ \cdots & \cdots \\ B_{1n} + k_{1n}(t - T_{1n}) & T_{1n} \le t < T/2 \end{cases} \tag{4.1}$$

$$f_{-}(t) = \begin{cases} k_{20}t & T_{21} \le t < 0 \\ B_{21} + k_{21}(t - T_{21}) & T_{22} \le t < T_{21} \\ \cdots & \cdots \\ B_{2n} + k_{2n}(t - T_{2n}) & -T/2 \le t < T_{2n} \end{cases} \tag{4.2}$$

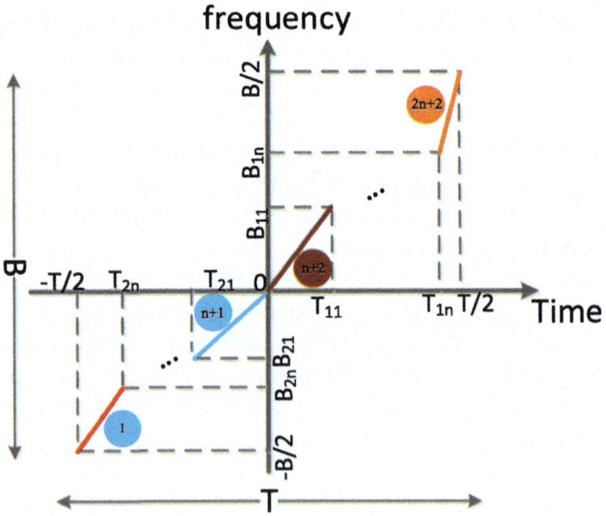

Fig. 4.1 2n + 2 stages for PWL functions

where B denotes the frequency bandwidth, T is the pulse duration, and k_{1i}, k_{2i} are the FM rates for each segment. The operators $[\cdot]+$ and $[\cdot]-$ correspond to the positive-axis and negative-axis components of the function, respectively.

$$k_{10} = \frac{B_{11}}{T_{11}}, \cdots, \underset{1 \le i < n}{k_{1i}} = \frac{B_{1(i+1)} - B_{1i}}{T_{1(i+1)} - T_{1i}}, \cdots, k_{1n} = \frac{\frac{B}{2} - B_{1n}}{\frac{T}{2} - T_{1n}}. \tag{4.3}$$

$$k_{20} = \frac{B_{21}}{T_{21}}, \cdots, \underset{1 \le i < n}{k_{2i}} = \frac{B_{2(i+1)} - B_{2i}}{T_{2(i+1)} - T_{2i}}, \cdots, k_{2n} = \frac{\frac{B}{2} + B_{2n}}{\frac{T}{2} + T_{2n}}. \tag{4.4}$$

The phase of this 2n + 2 stage NLFM waveform is then obtained by integrating (4.1) and (4.2) as follows:

$$\theta_{+}(t) = 2\pi \begin{cases} k_{10}t^2 + c_0 & 0 \le t < T_{11} \\ B_{11}t + k_{11}(t^2/2 - T_{11}t) + c_1 & T_{11} \le t < T_{12} \\ \cdots & \cdots \\ B_{1n} + k_{1n}(t^2/2 - T_{1n}t) + c_n & T_{1n} \le t < T/2 \end{cases} \tag{4.5}$$

$$\theta_{-}(t) = 2\pi \begin{cases} k_{20}t^2 + d_0 & T_{21} \le t < 0 \\ B_{21}t + k_{21}(t^2/2 - T_{21}t) + d_1 & T_{22} \le t < T_{21} \\ \cdots & \cdots \\ B_{2n} + k_{2n}(t^2/2 - T_{2n}t) + d_n & -T/2 \le t < T_{2n} \end{cases} \tag{4.6}$$

where c_i, d_i are constant

$$c_0 = 0, c_1 = -\frac{1}{2}B_{11}T_{11} + \frac{1}{2}k_{11}T_{11}^2 \tag{4.7}$$

$$\begin{aligned} \underset{2 \le i < n}{c_i} &= (B_{1(i-1)} - B_{1i})T_{1i} + \frac{1}{2}T_{1i}^2(k_{1(i-1)} + k_{1i}) \\ &\quad - k_{1(i-1)}T_{1i}T_{1(i-1)} + c_{i-1} \end{aligned} \tag{4.8}$$

$$d_0 = 0, d_1 = -\frac{1}{2}B_{21}T_{21} + \frac{1}{2}k_{21}T_{21}^2 \tag{4.9}$$

$$\begin{aligned} \underset{2 \le i < n}{d_i} &= (B_{2(i-1)} - B_{2i})T_{2i} + \frac{1}{2}T_{2i}^2(k_{2(i-1)} + k_{2i}) \\ &\quad - k_{2(i-1)}T_{2i}T_{2(i-1)} + d_{i-1} \end{aligned} \tag{4.10}$$

Each segment is assigned an equal duration, with $T_{1i} = \frac{T}{2n+2}i$ and $T_{2j} = -\frac{T}{2n+2}i$ for $(i = 1, 2, \ldots, n)$. A total of $2n$ parameters ($B_{11} \ldots B_{1n}$ and $B_{21} \ldots B_{2n}$) need to be determined to obtain the optimal NLFM waveform.

In this section, the focus is on the quasi-orthogonality of the signal, that is, minimizing the signal's CCE. At the same time, the waveform must maintain an acceptable peak sidelobe ratio (PSLR). This leads to a nonlinear constrained optimization problem:

$$\begin{aligned} &\min_{\vec{B}} f(\text{CCE}), \text{ such that} \\ &c(\text{PSLR}_{s_1}) \leq 0, c(\text{PSLR}_{s_2}) \leq 0, -B/2 \leq \vec{B} \leq B/2 \end{aligned} \tag{4.11}$$

where $\text{CCE} = \int |S_1(f)|^2 |S_2(f)|^2 df$ and $\vec{B} = (B_{11} \ldots B_{1n}, B_{21} \ldots B_{2n})$. The subscripts s_1 and s_2 refer to the transmitted baseband radar signals $s_1(t)$ and $s_2(t)$, with corresponding spectra $S_1(f)$ and $S_2(f)$. Additionally, PSLRs_1 and PSLRs_2 denote the peak sidelobe ratios of $s_1(t)$ and $s_2(t)$, respectively.

In this section, the ALGA is employed to handle the nonlinear constrained problem in (4.11). The ALGA can be regarded as an extended form of the Genetic Algorithm (GA) [30], which is frequently utilized for global optimization. The primary processing flow is shown in Fig. 4.2. Through our recent experiments, we found that improving the PSLR while keeping the main waveform (MW) unchanged typically introduces fluctuations in the signal spectrum. Similar observations have been reported in [31, 32]. These spectral variations, rather than a fully flat spectrum, are important for achieving reduced CCE, and Fig. 4.3 provides an intuitive illustration. In simple terms, after range matched filtering (MF), a spectrum with non-flat characteristics can be further shaped by another spectrum. Based on this understanding, the first step focuses on optimizing $s_1(t)$ so that a non-flat spectrum is deliberately produced. This process generates an improved NLFM signal with low sidelobes and a narrow main lobe, enabling the subsequent optimization steps to proceed more effectively. To construct a well-initialized quasi-orthogonal waveform $s_2(t)$, the spectrum of $s_1(t)$ is reversed, and the instantaneous frequency of $s_2(t)$ is defined as:

$$f_{s_2}(t) = -f_{s_1}(-t) \tag{4.12}$$

where $f_{s_1}(t)$ and $f_{s_2}(t)$ represent the instantaneous frequencies of signals $s_1(t)$ and $s_2(t)$, respectively. Finally, $s_1(t)$ is optimized by setting its initial instantaneous frequency to $f_{s_1}(t) = -f_{s_2}(-t)$. The optimization algorithm follows the same ALGA flowchart shown in Fig. 4.2. The details of ALGA processing are not the focus of this paper, and further information can be found in [33].

4.2.2 *Mitigating Range Ambiguities with Quasi-Orthogonal NLFM Waveforms*

The method for suppressing range ambiguities using quasi-orthogonal NLFM waveforms is similar to the approach employed in OFDM and up–down chirp signals.

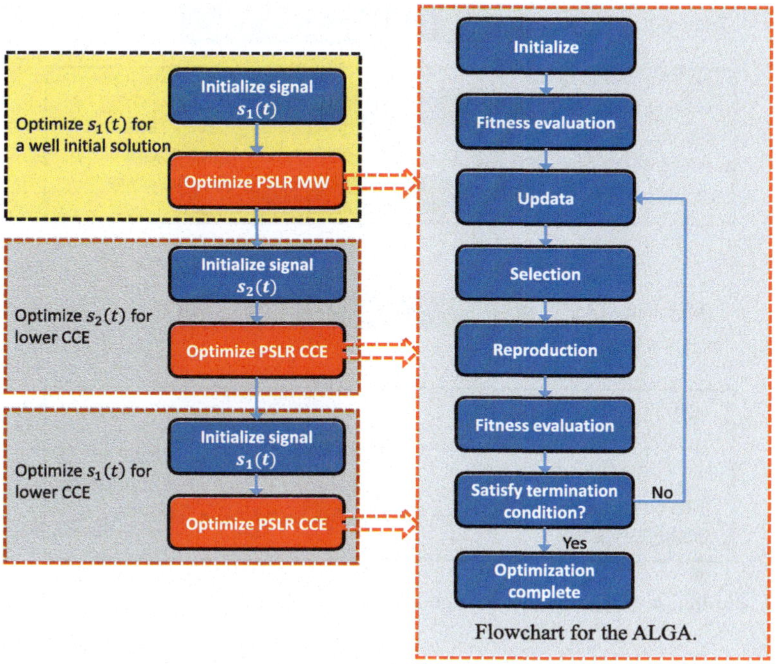

Fig. 4.2 The optimization process of orthogonal NLFM waveforms

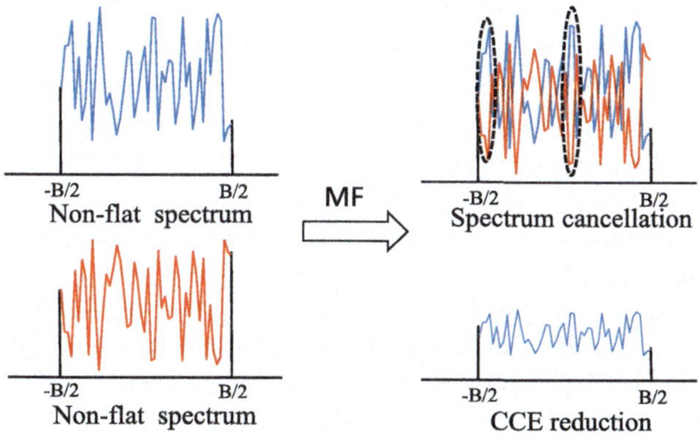

Fig. 4.3 Illustration of the reason why no-flat spectrum can reduce the CCE

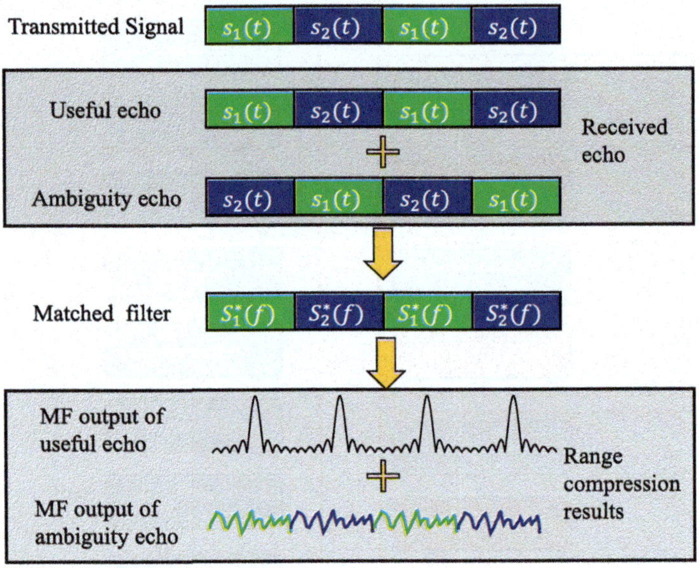

Fig. 4.4 Illustration of the concept of range-ambiguity suppression

For a point P within the scene, let its slant range be R_0, and define the slant range corresponding to the ith range ambiguity as:

$$R_{amb,i} = R_0 + i \times \text{PRI} \cdot c/2 \tag{4.13}$$

where PRI $= 1/\text{PRF}$ is the pulse repetition interval. Consequently, the ambiguity echo arrives at the antenna after i PRIs relative to the useful echo. In other words, even though the received signals overlap in time, the useful signal can be extracted if the transmitted pulses are orthogonal. This allows the range ambiguities to be effectively suppressed.

As an illustrative example (Fig. 4.4), two distinct NLFM waveforms ($s_1(t)$ and $s_2(t)$) are transmitted in an alternating manner. Relative to the useful echo, the $i = 1$th ambiguity echo is delayed by one PRI, which can be interpreted as the received waveform sequence lagging by one PRI. After range compression, the useful echo is properly matched, whereas the ambiguity echo becomes mismatched. Due to the orthogonality of the waveforms, the energy of the mismatched signal is suppressed, effectively reducing range ambiguities. It should be noted that if n orthogonal waveforms are transmitted, the ith ambiguity echo (where $i = k \times n$, $n = 1, 2, 3, \ldots$) cannot be suppressed because of the waveform's periodicity. Therefore, transmitting a larger number of orthogonal waveforms enables suppression of more ambiguity energy. However, considering the attenuation of antenna sidelobes, distant ambiguity echoes can typically be neglected.

4.3 Transmission Scheme and Signal Model

In this section, the signal model of the ATM is established, and a detailed performance evaluation, including the derivation of the RASR expressions, is provided.

4.3.1 Signal Model

In SAR systems, the ATM mode is implemented by cycling through a set of L distinct waveforms, with L defining the period over which the waveform sequence repeats. These waveforms, denoted as $s_1(t), s_2(t), \cdots, s_L(t)$, form the transmission pattern whose timing arrangement is depicted in Fig. 4.5.

Starting with the desired echo, consider a point target whose minimum slant range is R_0 and whose beam center is crossed at time η_c. For the waveform $s_l(t)$, where $l = 1, 2, \cdots, L$, the associated desired return can be written as [34]:

$$s_l(t, \eta) = \sigma_0 w_a(\eta - \eta_c)\, w_r\left(t - \frac{2R(\eta)}{c}\right) \times \exp\left(-j4\pi f_0 \frac{R(\eta)}{c}\right) \exp\left(j\theta_l\left(t - \frac{2R(\eta)}{c}\right)\right), \tag{4.14}$$

where σ_0 represents the backscatter coefficient of the desired echo. The variables t and η correspond to fast time and slow time, respectively. The functions $w_a(t)$ and $w_r(\eta)$ denote the envelopes along fast time and slow time. The carrier frequency is indicated by f_0, and c is the speed of light. The slant range as a function of slow time is $R(\eta)$, and $\theta_l(t)$ defines the phase of the transmitted waveform $s_l(t)$.

By applying range compression with $S_l^*(f_t)$ (where $S_l(f_t)$ denotes the Fourier transform of $s_l(t)$), the resulting compressed echo $s_l(t, \eta)$ can be written as

$$s_l^{\mathrm{rc}}(t, \eta) = \sigma_0 w_a(\eta - \eta_c) \times \exp\left(-j4\pi f_0 \frac{R(\eta)}{c}\right) p_l\left(t - \frac{2R(\eta)}{c}\right), \tag{4.15}$$

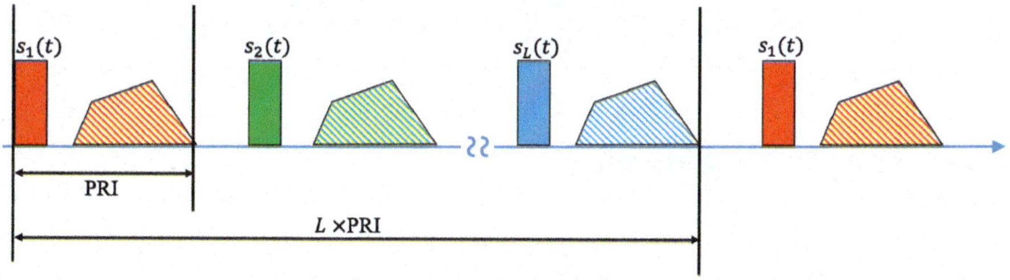

Fig. 4.5 Timing diagram of the ATM

where $p_l(t)$ denotes the output of the matched filter corresponding to the transmitted waveform $s_l(t)$, that is,

$$p_l(t) = \mathscr{F}_t^{-1}\left\{S_l(f_t)S_l^*(f_t)\right\}. \tag{4.16}$$

Accordingly, the desired echo signal following azimuth sampling can be expressed as

$$\hat{s}(t,\eta) = \sum_{l=1}^{L}\sum_{n}\delta\left(\eta - \frac{Ln}{f_a} - \frac{l-1}{f_a}\right)s_l^{\mathrm{rc}}(t,\eta), \tag{4.17}$$

where f_a is the PRF.

Next, considering the range ambiguity, the kth ambiguous signal in range can be expressed as

$$\hat{s}_k(t,\eta) = \sum_{l=1}^{L}\sum_{n}\delta\left(\eta - \frac{Ln}{f_a} - \frac{l-1}{f_a}\right)s_{l-k,l}^{\mathrm{rc}}(t,\eta), \tag{4.18}$$

where

$$\begin{aligned} s_{l-k,l}^{\mathrm{rc}}(t,\eta) &= \sigma_0^k w_a(\eta-\eta_c)\, w_r\left(t - \frac{2R_k(\eta)}{c}\right) \\ &\times \exp\left(-j4\pi f_0\frac{R_k(\eta)}{c}\right) p_{l-k,l}\left(t - \frac{2R_k(\eta)}{c}\right), \end{aligned} \tag{4.19}$$

where σ_0^k and $R_k(\eta)$ represent the backscatter coefficient and the slant range trajectory of the kth ambiguous echo, respectively. The term $p_{l-k,l}(t)$ corresponds to the output of the mismatched filtering between $S_{l-k}(f_t)$, which is the waveform associated with the kth ambiguity, and $S_l^*(f_t)$, the matched filter of the lth transmitted waveform, i.e.,

$$p_{l-k,l}(t) = \mathscr{F}_t^{-1}\left\{S_{l-k}(f_t)S_l^*(f_t)\right\}. \tag{4.20}$$

Furthermore, it should be noted that the subscripts $l-k$ in (4.18) and (4.19) are determined as follows

$$l-k = \begin{cases} l-k(\mathrm{mod}\ L), & l-k(\mathrm{mod}\ L) > 0 \\ L+l-k(\mathrm{mod}\ L), & l-k(\mathrm{mod}\ L) \le 0 \end{cases}. \tag{4.21}$$

4.3.2 *Imaging Performance*

In this subsection, we provide a detailed analysis of the imaging characteristics of ATM. Based on (4.17), the signal expression after applying Range Cell Migration Correction (RCMC) in the Range–Doppler (RD) domain can be formulated as

$$\hat{S}(t, f_\eta) = \frac{f_a}{L}\sum_n \sum_{l=1}^{L} S_l^{\mathrm{rc,rcmc}}\left(t, f_\eta - n\frac{f_a}{L}\right) \times \exp\left(-j2\pi n\frac{l-1}{L}\right), \tag{4.22}$$

where $S_l^{\mathrm{rc,\,rcmc}}(t, f_\eta)$ denotes the echo signal in the RD domain after range compression and RCMC corresponding to the transmitted waveform $s_l(t)$, which can be expressed as [34]

$$\begin{aligned} S_l^{\mathrm{rc,rcmc}}(t, f_\eta) &= \sigma_0 W_a\left(f_\eta - f_{\eta_c}\right) \\ &\times \exp\left(-j4\pi f_0 \frac{R_{\mathrm{rd}}(f_\eta)}{c}\right) p_l\left(t - \frac{2R_0}{c}\right) \\ &= S_a(f_\eta) p_l\left(t - \frac{2R_0}{c}\right), \end{aligned} \tag{4.23}$$

where $R_{\mathrm{rd}}(f_\eta)$ denotes the range evolution in the RD domain, and f_η together with f_{η_c} represent the Doppler frequency and its center, respectively. The term $W_a(f_\eta)$ specifies the Doppler envelope, while $S_a(f_\eta)$ refers to the common azimuth-spectrum component shared across different waveforms, i.e.,

$$S_a(f_\eta) = \sigma_0 W_a\left(f_\eta - f_{\eta_c}\right) \exp\left(-j4\pi f_0 \frac{R_{\mathrm{rd}}(f_\eta)}{c}\right). \tag{4.24}$$

For simplicity, the explicit mathematical expressions of $R_{\mathrm{rd}}(f_\eta)$ and f_{η_c} are omitted, and the interested reader can refer to Sect. 5 of [34] for details. In addition, (4.22) can be rewritten as

$$\hat{S}(t, f_\eta) = \frac{f_a}{L}\sum_{l=1}^{L} p_l\left(t - \frac{2R_0}{c}\right) \times \sum_n S_a\left(f_\eta - n\frac{f_a}{L}\right) \exp\left(-j2\pi n\frac{l-1}{L}\right). \tag{4.25}$$

The equation above indicates that using diverse waveforms with distinct autocorrelation functions leads to azimuth spectrum aliasing, which introduces additional azimuth ambiguities. Conversely, if all transmitted waveforms share the same autocorrelation function, i.e., $p_l(t) = p(t)$, then (4.25) can be further written as

$$\hat{S}(t, f_\eta) = \frac{f_a}{L} p\left(t - \frac{2R_0}{c}\right) \times \sum_n S_a\left(f_\eta - n\frac{f_a}{L}\right) \sum_{l=1}^{L} \exp\left(-j2\pi n\frac{l-1}{L}\right). \tag{4.26}$$

Noting that

$$\sum_{l=1}^{L} \exp\left(-j2\pi n\frac{l-1}{L}\right) = \begin{cases} L, & n=0 \\ 0, & \text{elsewhere} \end{cases}. \tag{4.27}$$

Therefore, (4.26) can be reformulated as

$$\hat{S}(t, f_\eta) = f_a S_a\left(f_\eta\right) p\left(t - \frac{2R_0}{c}\right), \tag{4.28}$$

which shows that employing waveforms with identical auto-correlation properties avoids introducing additional azimuth ambiguities, whereas the range ambiguities remain since their power spectral densities are the same. Furthermore, the peak responses of different waveforms coincide, ensuring that no azimuth-spectrum aliasing appears at the peak range cell. These observations imply that azimuth-ambiguity compensation within the main-lobe range support is still necessary, and the corresponding approach is detailed in Sect. 4.4.

4.3.3 Ambiguity Suppression Performance

In this subsection, the performance of range suppression is presented along with the corresponding RASR equations.

The RASR is defined as the ratio of the power of the range-ambiguous echo to that of the primary return and is usually estimated as [5]

$$\text{RASR} \cong \frac{\sum_k \frac{\sigma_0(\phi_k) G_r^2(\varphi_k)}{R_k^3 \sin(\phi_k)}}{\frac{\sigma_0(\phi_0) G_r^2(\varphi_k)}{R_0^3 \sin(\phi_0)}} \tag{4.29}$$

where $\sigma_0(\phi)$ denotes the backscatter coefficient as a function of the incidence angle ϕ, while $G_r(\varphi)$ represents the two-way antenna gain pattern in elevation with respect to the look angle φ. The angles ϕ_k and φ_k (ϕ_0 and φ_0) refer to the incidence and look angles of the kth range-ambiguous signal (and the desired echo), respectively. Moreover, R_0 and R_k correspond to the slant ranges of the desired return and the kth range ambiguity.

Equation (4.29) relies on the assumption that, up to a scaling factor, the azimuth pattern remains identical for all elevation angles. While this condition is generally satisfied for planar antennas, it may lead to inaccuracies when applied to reflector antennas, whose two-dimensional radiation pattern cannot be decomposed into separate azimuth and elevation components [35]. To address this limitation, [35] introduces a more general RASR formulation by modeling the two-way antenna power pattern as a joint function of the look angle ϕ_k and the Doppler frequency f_η.

$$\text{RASR} \cong \frac{\dfrac{\sum_k \sigma_0(\phi_k) \int_{-B_a/2}^{B_a/2} G_r^2(\varphi_k, f_\eta)}{R_k^3 \sin(\phi_k)}}{\dfrac{\sigma_0(\phi_0) \int_{-B_a/2}^{B_a/2} G_r^2(\varphi_0, f_\eta)}{R_0^3 \sin(\phi_0)}}, \tag{4.30}$$

where B_a denotes the processed Doppler bandwidth. According to (4.19), the kth range-ambiguous signal ($k \neq iL, i = \pm 1, \pm 2, \dots$) becomes mismatched after range compression. As a result, the ambiguity power is reduced due to the low cross-correlation energy (CCE), which demonstrates the mechanism of range ambiguity suppression using ATM combined with waveform diversity [20]. Based on this, the RASR can be estimated as[1]

$$\text{RASR} \cong \sum_{l=1}^{L} \frac{\dfrac{\sum_k \sigma_0(\phi_k) \int_{-T}^{T} |p_{l-k,l}(t)|^2 dt \int_{-B_a/2}^{B_a/2} G_r^2(\varphi_k, f_\eta)}{R_k^3 \sin(\phi_k)}}{\dfrac{\sigma_0(\phi_0) |p_{l,l}(0)|^2 dt \int_{-B_a/2}^{B_a/2} G_r^2(\varphi_0, f_\eta)}{R_0^3 \sin(\phi_0)}} \tag{4.31}$$

It should be emphasized that, regarding (4.31), in a scene with distributed targets, the energy from range ambiguities across all range cells must be taken into account, which requires the full-time integration shown in (4.31). At the same time, (4.31) evaluates the RASR after range compression, where only the peak energy of the desired signal contributes to the effective output. This is the reason why the denominator includes the term $|p_{l,l}(0)|^2$.

4.4 Signal Processing

In this section, a novel azimuth compensation method is presented, which utilizes *a priori* knowledge of the transmitted waveforms.

Expressing the signal in vector form facilitates the subsequent analysis, since the compensation algorithm is implemented in the digital domain. After range compression and RCMC, the received signal can be written as[2]

$$r(n) = \sigma_0 \sum_{l=1}^{L} p_l(\bar{\tau}) z_l(n) s(n), n = 0, 1, \cdots, N-1, \tag{4.32}$$

where $\boldsymbol{r} = [r(0), r(1), \cdots, r(N-1)]$ represents the received signal, $\bar{\tau}$ denotes an unknown time cell within the range main-lobe interval $[-\tau_m, \tau_m]$, $p_l(\bar{\tau})$ is the auto-

[1] Both the desired and range-ambiguous returns undergo a Doppler shift induced by waveform diversity. While the proposed method compensates for the Doppler shift of the desired signal (see Sect. 4.3), the range-ambiguous echoes remain shifted, enabling further suppression via the azimuth filter. Consequently, the actual RASR is expected to be lower than the estimation given in (4.31).

[2] Contributions from distant targets are negligible because of the low sidelobes of the designed waveforms.

correlation of the lth transmitted waveform, and $\boldsymbol{s} = [s(0), s(1), \cdots, s(N-1)]$ corresponds to the original azimuth-modulated signal.

$$z_l(n) = \begin{cases} 1, & n = l + i \times L - 1, i = 0, 1, 2, \cdots \\ 0, & \text{elsewhere} \end{cases}. \tag{4.33}$$

Furthermore, $z_l(n)$ can be recast as (see Appendix 4.7)

$$z_l(n) = \frac{1}{L} \sum_{l'=0}^{L-1} e^{-j\left(\frac{2\pi}{L}\right)(n-l+1)l'}. \tag{4.34}$$

The discrete spectrum of $\boldsymbol{r}$ can be represented as

$$\begin{aligned} R(k) &= \sum_{n=0}^{N-1} r(n) e^{-j\left(\frac{2\pi}{N}\right)kn} \\ &= \sigma_0 \sum_{n=0}^{N-1} \sum_{l=1}^{L} p_l(\bar{\tau}) z_l(n) s(n) e^{-j\left(\frac{2\pi}{N}\right)kn} \\ &= \sigma_0 \sum_{l=1}^{L} p_l(\bar{\tau}) \sum_{n=0}^{N-1} \left(\frac{1}{L} \sum_{l'=0}^{L-1} e^{-j\left(\frac{2\pi}{L}\right)(n-l+1)l'} \right) s(n) e^{-j\left(\frac{2\pi}{N}\right)kn} \\ &= \frac{\sigma_0}{L} \sum_{l=1}^{L} p_l(\bar{\tau}) \sum_{n=0}^{N-1} \sum_{l'=0}^{L-1} e^{j\left(\frac{2\pi}{L}\right)(l-1)l'} s(n) e^{-j\left(\frac{2\pi}{N}\right)\left(k+\frac{Nl'}{L}\right)n} \\ &= \frac{\sigma_0}{L} \sum_{l=1}^{L} p_l(\bar{\tau}) \sum_{l'=0}^{L-1} e^{j\left(\frac{2\pi}{L}\right)(l-1)l'} S\left(k + \frac{Nl'}{L}\right), \end{aligned} \tag{4.35}$$

where $S(k), k = 0, 1, \cdots, N-1$ denotes the discrete spectrum of $\boldsymbol{s}$. Equation (4.35) further reveals that variations in the auto-correlation values, i.e., $p_l(\bar{\tau})$, induce aliasing in the azimuth spectrum.

The right portion of the main-lobe, i.e., $[0, \tau_m]$, is partitioned into M range cells, based on which the azimuth compensation vector $\boldsymbol{g}_m = [g_m(0), g_m(1), \cdots, g_m(N-1)], ; m = 0, \cdots, M-1$, can be formulated as

$$g_m(n) = \sum_{l=1}^{L} \frac{1}{p_l\left(\frac{m\tau_m}{M}\right)} z_l(n), n = 0, 1, \cdots, N-1. \tag{4.36}$$

Normalizing $\boldsymbol{g}_m$, as

$$\boldsymbol{g}_m = \frac{\boldsymbol{g}_m}{\|\boldsymbol{g}_m\|_2}. \tag{4.37}$$

Table 4.1 Azimuth compensation algorithm

- Constructing compensation vectors according to (4.36) and (4.37)
- The optimal estimate of $\bar{\tau}$ is

$$\hat{\bar{\tau}} = \frac{\hat{m}\tau_m}{M}, \tag{4.40}$$

where

$$\hat{m} = \max_{m} |\hat{R}_m(k = k_0)|, \tag{4.41}$$

where $\hat{R}_m(k)$, $k = 0, \cdots, N-1$ is the spectrum of $\hat{\boldsymbol{r}}_m$, and k_0 refers to the index corresponding to Doppler center position.
- Selecting the optimal compensation vector $\boldsymbol{g}_{\hat{m}}$, and performing azimuth compensation, as

$$\frac{\|\boldsymbol{r}\|_2}{\left\|\boldsymbol{r} \odot \boldsymbol{g}_{\hat{m}}\right\|_2}\boldsymbol{r} \odot \boldsymbol{g}_{\hat{m}} \tag{4.42}$$

The compensated signal can be expressed as

$$\hat{\boldsymbol{r}}_m = \boldsymbol{r} \odot \boldsymbol{g}_m, \tag{4.38}$$

where $\hat{\boldsymbol{r}}_m = [\hat{r}_m(0), \cdots, \hat{r}_m(N-1)]$ denotes the signal vector compensated by $\boldsymbol{g}_m$.
Normalizing $\hat{\boldsymbol{r}}_m$, as

$$\hat{\boldsymbol{r}}_m = \frac{\|\boldsymbol{r}\|_2}{\left\|\hat{\boldsymbol{r}}_m\right\|_2}\hat{\boldsymbol{r}}_m \tag{4.39}$$

Applying the optimal compensation vector maximizes the spectrum amplitude at the Doppler center, as the aliasing spectrum energy is minimized. Following this criterion, the compensation algorithm is presented in Table 4.1.

4.5 Simulation Experiment

This section conducts detailed simulation studies based on the LT-1 system parameters given in Table 4.2 to examine the design of quasi-orthogonal waveforms and assess the imaging performance of ATM.

4.5.1 Point Target Simulation

Initially, following the quasi-orthogonal NLFM waveform design approach described in [25], the vector $\boldsymbol{b}_{\text{Initial}}$ with $N = 130$ is generated by sampling the instantaneous

Table 4.2 Main simulation parameters of the LT-1 system

Parameters	Value
Orbit height	607 Km
Look angle access range	24.59–26.84 deg
SAR pulse width	60 μs
Bandwidth	60 MHz
PRF	1723 Hz
Antenna type	Planar

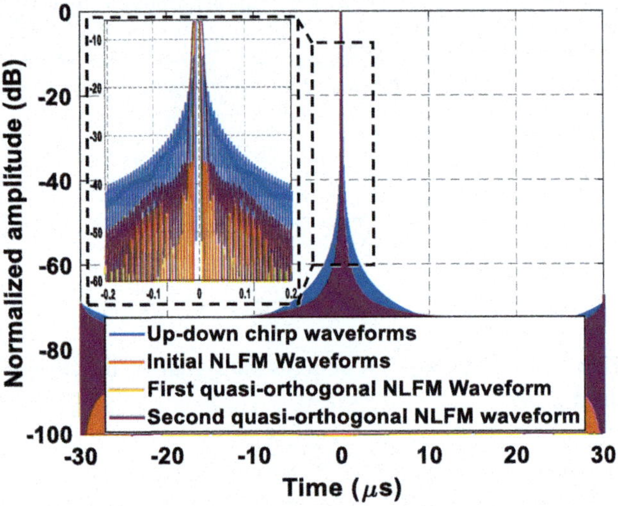

Fig. 4.6 Matched filter outputs in dB of different quasi-orthogonal waveforms

frequency of an NLFM waveform whose spectrum approximates a Taylor window with a peak sidelobe level of −35 dB. The PSLR threshold is set to −35 dB, and the algorithm stops when $\delta = 10^{-6}$. Figure 4.6 shows the matched-filter responses for the initial NLFM waveform, the optimized quasi-orthogonal NLFM, and the up-down chirp signals, while the corresponding cross-correlation functions are presented in Fig. 4.7. Key performance metrics for these waveform pairs are listed in Table 4.3. The results indicate that the optimized NLFM signals achieve lower CCE compared with other waveforms, while maintaining PSLR and ISLR values similar to conventional imaging without windowing, thereby avoiding any loss in SNR.

Subsequently, simulations are performed using a point target located at midrange[3] to evaluate the imaging performance of ATM. The optimized quasi-orthogonal NLFM waveforms shown in Fig. 4.6 are employed for this assessment. The resulting

[3] The point target is positioned at $(x, y) = (351, 3201)$, where x and y represent the azimuth and range sample indices, respectively. The total number of range and azimuth samples are 6401 and 701.

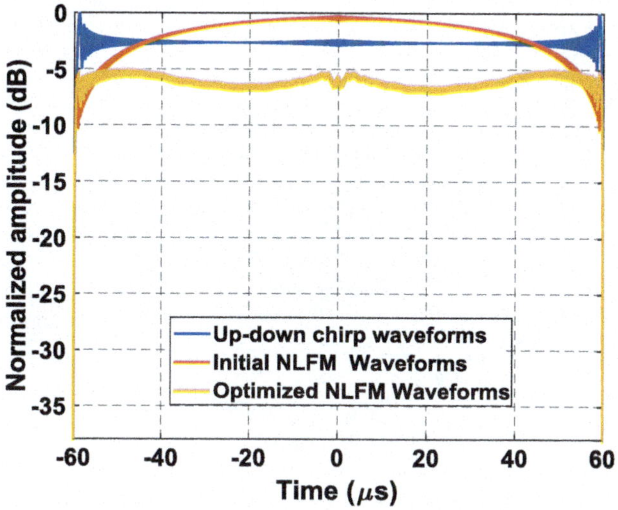

Fig. 4.7 Cross-correlation responses in dB for the different quasi-orthogonal waveforms

Table 4.3 Performance metrics for the different waveforms

Waveforms	PSLR (dB)	ISLR (dB)	3-dB MW (sample)	CCE[a] (dB)
Up-down chirp	−13.3	−9.7	1	0
Initial NLFM	−35	−26.6	1.3	0.91
First quasi-orthogonal NLFM	−35.8	−27.6	1.57	−3.5
Second quasi-orthogonal NLFM	−35.6	−27.6	1.56	

[a] The CCE relative to a chirp signal

focused image is presented in Fig. 4.8a, where additional azimuth ambiguities arising from waveform diversity can be observed. Azimuth profiles at several range positions within the main-lobe support are depicted in Fig. 4.8b and c. As discussed in Sect. 4.3.2, the matched-filter peak responses are the same for all waveforms; therefore, the azimuth profile at the peak range cell (the 3201st cell), shown in Fig. 4.8b, demonstrates proper focusing. In comparison, the adjacent range cell (3202nd), illustrated in Fig. 4.8c, exhibits significant azimuth ambiguities. After implementing the proposed compensation method, the corrected results in Fig. 4.9 show that the azimuth signals within the main-lobe range are effectively restored and the ambiguities introduced by waveform diversity are successfully removed.

Finally, to evaluate the interference at the first near-ambiguous range caused by a midrange point target, Fig. 4.10 shows the imaging results corresponding to different waveforms of the range-ambiguous echo. In these simulations, all raw signals are

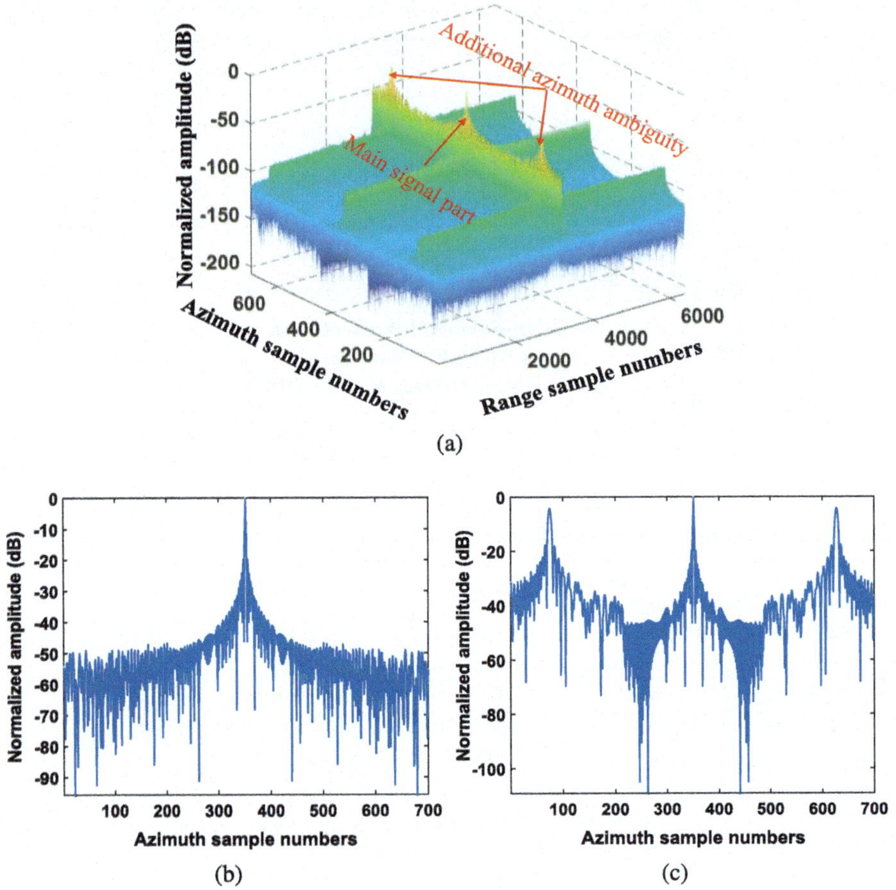

Fig. 4.8 Focused results for a midrange point target. **a** Focused image. **b** Azimuth profile at range cell 3201. **c** Azimuth profile at range cell 3202

assumed to have the same energy, and the influences of antenna gain, backscatter coefficient, incidence angle, and slant range are neglected. The ambiguity suppression rates for different transmission schemes, defined as the total ambiguity energy relative to a standard up-chirp waveform, are listed in Table 4.4. As illustrated in Fig. 4.10b, c, and d, residual ambiguities appear along the range dimension when quasi-orthogonal waveforms are applied. In addition, the range-ambiguous signals experience Doppler shifts due to differences in their cross-correlation functions. Consequently, further suppression of range ambiguities can be achieved via azimuth filtering, and the observed ambiguity suppression rates (Table 4.4) are lower than the reductions in CCE shown in Table 4.3. Overall, the optimized quasi-orthogonal NLFM waveforms provide superior range-ambiguity suppression performance due to their minimal CCE compared with other quasi-orthogonal waveform designs.

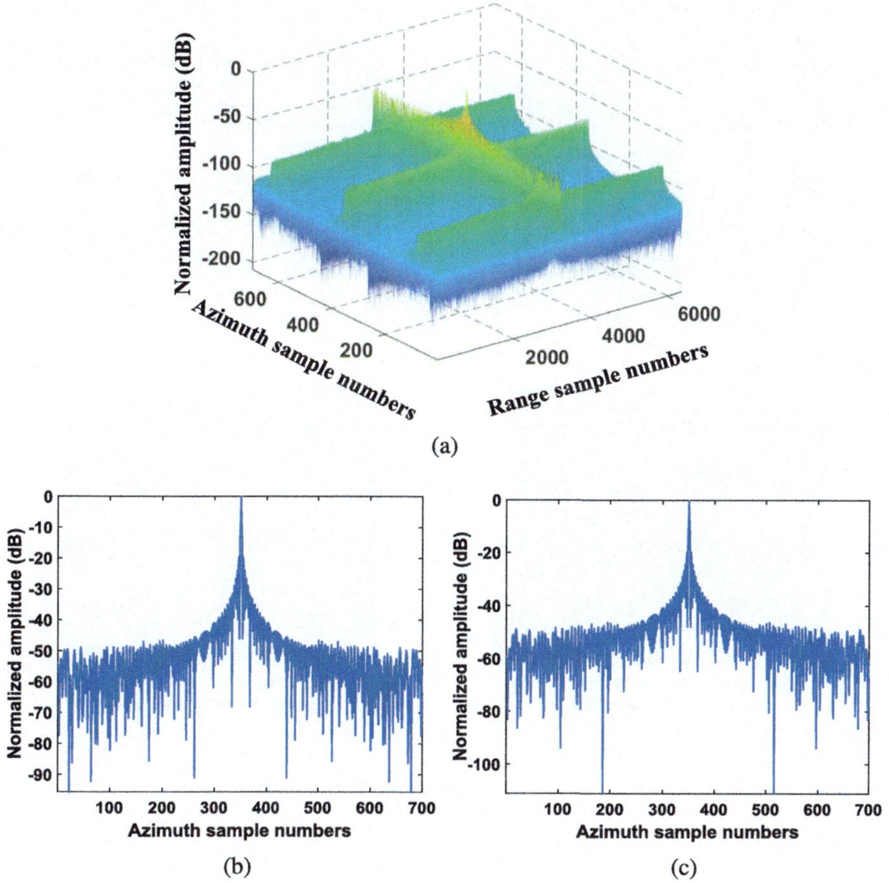

Fig. 4.9 Focused results following azimuth compensation. **a** Focused image. **b** Azimuth profile at range cell 3201. **c** Azimuth profile at range cell 3202

4.5.2 Distributed Target Simulation

In this subsection, distributed target experiments are conducted to further assess ambiguity interference effects, based on the LT-1 system parameters listed in Table 4.2.

The simulation results for the first near-range ambiguity under different transmission schemes are shown in Fig. 4.11, where the horizontal axis corresponds to range and the vertical axis to azimuth. Figure 4.11a and b illustrate the desired and ambiguity regions, respectively, while Fig. 4.11c–f present the imaging results for the various transmission schemes. As indicated by the yellow rectangles, the optimized quasi-orthogonal NLFM waveforms achieve the most effective suppression of range ambiguities. Furthermore, Fig. 4.12 depicts the azimuth profiles of range ambiguity energy for different waveforms, demonstrating that the optimized NLFM signals

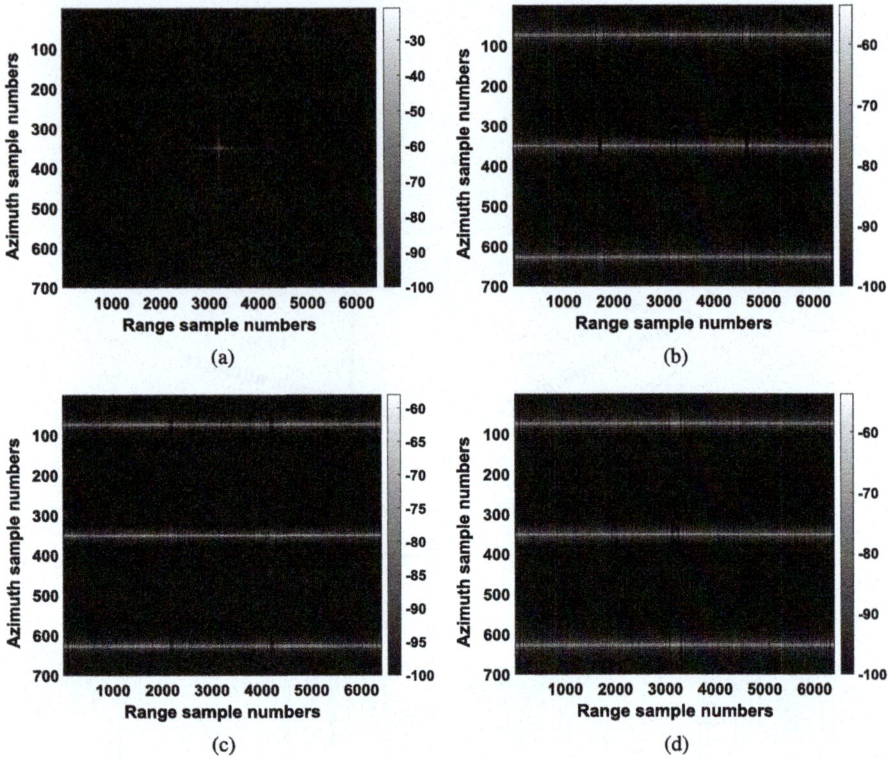

Fig. 4.10 Focused results at the first near-range ambiguity region of the image for: **a** up-chirp waveform, **b** up-down chirp waveforms, **c** initial quasi-orthogonal NLFM waveforms, and **d** optimized quasi-orthogonal NLFM waveforms.

Table 4.4 Range ambiguity suppression performance of a point target

Transmitted waveforms	Ambiguity suppression rate (dB)
Up chirp	0
Up-down chirp	−1.22
Initial NLFM	−0.35
Optimized quasi-orthogonal NLFM	−4.63

exhibit significantly lower energy than the alternatives. The range-suppression performance for each scheme is summarized in Table 4.5, which aligns well with the results from the point target simulations shown in Table 4.4.

(a)

(b)

(c)

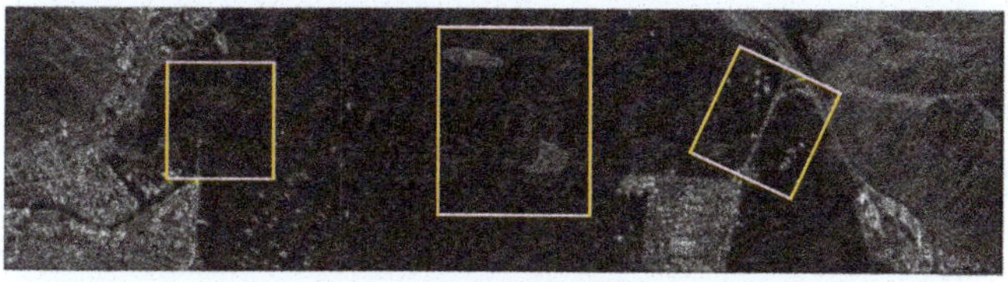

(d)

(e)

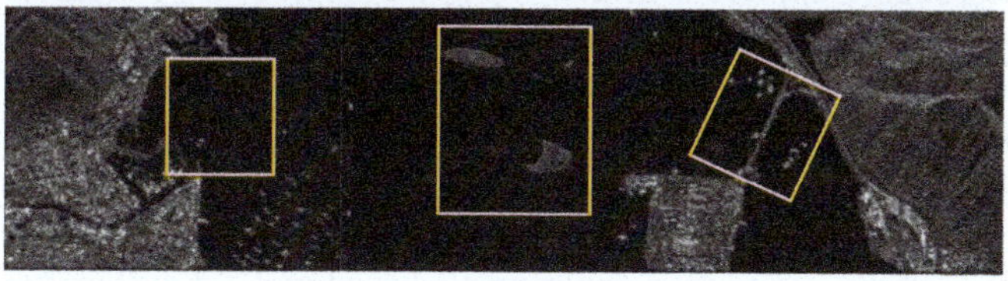

(f)

Fig. 4.11 Distributed scene simulation for the first near-range ambiguity region. **a** Image of the desired area. **b** Image of the range-ambiguity region. Imaging results for: **c** up-chirp waveform, **d** initial NLFM waveforms, **e** up-down chirp waveforms, and **f** optimized quasi-orthogonal NLFM waveforms

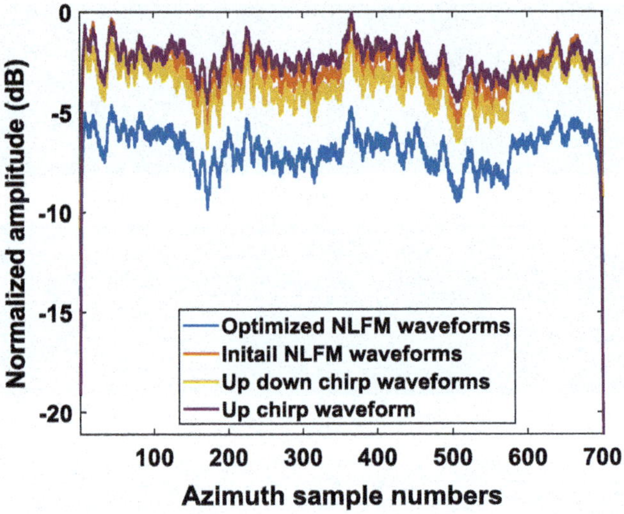

Fig. 4.12 Azimuth profiles of the range-averaged energy corresponding to the range ambiguities

Table 4.5 Range ambiguity suppression performance of distributed targets

Transmitted waveforms	Ambiguity suppression rate (dB)
Up chirp	0
Up-down chirp	−1.23
Initial NLFM	−0.36
Optimized quasi-orthogonal NLFM	−4.65

4.6 Summary

ATM has emerged as an effective approach for range ambiguity suppression because of its low implementation complexity and the additional design flexibility it provides. Previous studies have primarily focused on the design of quasi-orthogonal waveforms, while the impact of waveform diversity on imaging performance has been less thoroughly explored. In this work, a comprehensive signal model and a systematic processing framework for ATM are developed. Differences in the autocorrelation characteristics of the transmitted waveforms can introduce extra azimuth ambiguities; to address this, a new azimuth compensation method that leverages *a priori* knowledge of the transmitted signals is proposed. Furthermore, the integration of quasi-orthogonal NLFM waveforms within ATM is discussed to satisfy system requirements. Finally, simulation experiments based on LT-1 system parameters are carried out to validate the theoretical analysis.

4.7 The Derivation of (4.34)

If $n \neq l + i \times L - 1$, (4.34) can be further simplified as

$$
\begin{aligned}
z_l(n) &= \frac{1}{L}\sum_{l'=0}^{L-1} e^{-j\left(\frac{2\pi}{L}\right)(n-l+1)l'} \\
&= \frac{1}{L}\left(\frac{1 - e^{-j\left(\frac{2\pi}{L}\right)(n-l+1)L}}{1 - e^{-j\left(\frac{2\pi}{L}\right)(n-l+1)}}\right) \\
&= \frac{1}{L}\left(\frac{1 - 1}{1 - e^{-j\left(\frac{2\pi}{L}\right)(n-l+1)}}\right) = 0.
\end{aligned}
\tag{4.43}
$$

Otherwise, if $n = l + i \times L - 1$,

$$
\begin{aligned}
z_l(n) &= \frac{1}{L}\sum_{l'=0}^{L-1} e^{-j\left(\frac{2\pi}{L}\right)(n-l+1)l'} \\
&= \frac{1}{L}\sum_{l'=0}^{L-1} e^{-j\left(\frac{2\pi}{L}\right)iLl'} = \frac{1}{L}\sum_{l'=0}^{L-1} 1 = 1
\end{aligned}
\tag{4.44}
$$

Therefore, (4.34) satisfies the definition of function $z_l(n)$, i.e., (4.33).

References

1. Werninghaus R, Buckreuss S. The TerraSAR-X Mission and System Design. IEEE Transactions on Geoscience and Remote Sensing. 2010;48(2):606–614.
2. Caltagirone F, Luca GD, Covello F, et al. Status, results, potentiality and evolution of COSMO-SkyMed, the Italian Earth Observation constellation for risk management and security. In: 2010 IEEE International Geoscience and Remote Sensing Symposium; 2010. p. 4393–4396.
3. Dierking W. Mapping of Different Sea Ice Regimes Using Images From Sentinel-1 and ALOS Synthetic Aperture Radar. IEEE Transactions on Geoscience and Remote Sensing. 2010 ;48(3):1045–1058.
4. Deng YK, Zhao FJ, Wang Y. Brief analysis on the development and application of spaceborne SAR. Journal of Radars. 2012;1(1):1–10.
5. Curlander JC, Mcdonough RN. Synthetic Aperture Radar: Systems and Signal Processing. New York, NY, USA: Wiley; 1991.
6. Krieger G, Gebert N, Moreira A. Multidimensional Waveform Encoding: A New Digital Beamforming Technique for Synthetic Aperture Radar Remote Sensing. IEEE Transactions on Geoscience and Remote Sensing. 2008 ;46(1):31–46.
7. Krieger G, Gebert N, Younis M, et al. Advanced synthetic aperture radar based on digital beamforming and waveform diversity. In: 2008 IEEE Radar Conference; 2008. p. 1–6.
8. Liu M, Zhang B, Xu Z, et al. Ambiguities Suppression for Azimuth Multichannel SAR Based on $L_{2,q}$ Regularization With Application to Gaofen-3 Ultra-Fine Stripmap Mode. IEEE Journal of Selected Topics in Applied Earth Observations and Remote Sensing. 2021;14:1532–1544.

9. Krieger G. MIMO-SAR: Opportunities and Pitfalls. IEEE Transactions on Geoscience and Remote Sensing. 2014;52(5):2628–2645.
10. Jin G, Deng Y, Wang W, et al. Segmented Phase Code Waveforms: A Novel Radar Waveform for Spaceborne MIMO-SAR. IEEE Transactions on Geoscience and Remote Sensing. 2020;p. 1–16.
11. Lan L, Xu J, Liao G, et al. Suppression of Mainbeam Deceptive Jammer With FDA-MIMO Radar. IEEE Transactions on Vehicular Technology. 2020;69(10):11584–11598.
12. Lan L, Liao G, Xu J, et al. Transceive Beamforming With Accurate Nulling in FDA-MIMO Radar for Imaging. IEEE Transactions on Geoscience and Remote Sensing. 2020;58(6):4145–4159.
13. Zhang Y, Wang W, Deng Y, et al. Quadratically Constrained Ambiguity Suppression Algorithm for APC/Multichannel SAR Systems With Nonuniform Spatial Sampling. IEEE Transactions on Geoscience and Remote Sensing. 2021;59(2):1319–1330.
14. Dall J, Kusk A. Azimuth phase coding for range ambiguity suppression in SAR. In: IGARSS 2004. 2004 IEEE International Geoscience and Remote Sensing Symposium. vol. 3; 2004. p. 1734–1737 vol.3.
15. Bordoni F, Younis M, Krieger G. Ambiguity Suppression by Azimuth Phase Coding in Multichannel SAR Systems. IEEE Transactions on Geoscience and Remote Sensing. 2012;50(2):617–629.
16. Villano M, Krieger G, Moreira A. Nadir Echo Removal in Synthetic Aperture Radar via Waveform Diversity and Dual-Focus Postprocessing. IEEE Geoscience and Remote Sensing Letters. 2018;15(5):719–723.
17. Jin G, Wang W, Deng Y, et al. A Novel Range-Azimuth Joint Modulation Scheme for Range Ambiguity Suppression. IEEE Transactions on Geoscience and Remote Sensing. 2021;p. 1–10.
18. Callaghan GD, Longstaff ID. Wide-swath space-borne SAR and range ambiguity. In: Radar 97 (Conf. Publ. No. 449); 1997. p. 248–252.
19. Riché V, Méric S, Baudais J, et al. Investigations on OFDM Signal for Range Ambiguity Suppression in SAR Configuration. IEEE Transactions on Geoscience and Remote Sensing. 2014;52(7):4194–4197.
20. Wang W. Mitigating Range Ambiguities in High-PRF SAR With OFDM Waveform Diversity. IEEE Geoscience and Remote Sensing Letters. 2013 ;10(1):101–105.
21. Xu Zh, Wang R, Ye K, et al. Simultaneous range ambiguity mitigation and sidelobe reduction using orthogonal non-linear frequency modulated (ONLFM) signals for satellite SAR Imaging. Remote Sensing Letters. 2018;9(9):829–838.
22. Mittermayer J, Martinez JM. Analysis of range ambiguity suppression in SAR by up and down chirp modulation for point and distributed targets. In: IGARSS 2003. 2003 IEEE International Geoscience and Remote Sensing Symposium.. vol. 6; 2003. p. 4077–4079.
23. Xu W, Huang P, Tan W. Azimuth Phase Coding by Up and Down Chirp Modulation for Range Ambiguity Suppression. IEEE Access. 2019;7:143780–143791.
24. Knapskog AO. Range Ambiguity Suppression in Space-Borne SAR by Up- and Down-Chirp Modulation in Combination with Pseudo-Random Bi-Phase Coding. In: EUSAR 2018; 12th European Conference on Synthetic Aperture Radar; 2018. p. 1–4.
25. Jin G, Aubry A, De Maio A, et al. Quasi-Orthogonal Waveforms for Ambiguity Suppression in Spaceborne Quad-Pol SAR. IEEE Transactions on Geoscience and Remote Sensing. 2021;p. 1–17.
26. Jin G, Deng Y, Wang W, et al. On the SAR Imaging Performance Analysis of Alternate Transmitting Mode Based on Waveform Diversity: Theory and Simulation. IEEE Geoscience and Remote Sensing Letters. 2020;17(9):1553–1557.
27. Jin G, Liu K, Liu D, et al. An Advanced Phase Synchronization Scheme for LT-1. IEEE Transactions on Geoscience and Remote Sensing. 2020;58(3):1735–1746.
28. Kurdzo JM, Cheong BL, Palmer RD, et al. Optimized NLFM pulse compression waveforms for high-sensitivity radar observations. In: 2014 International Radar Conference; 2014. p. 1–6.
29. Saeedi J, Faez K. Synthetic aperture radar imaging using nonlinear frequency modulation signal. IEEE Transactions on Aerospace and Electronic Systems. 2016 ;52(1):99–110.

30. Holland JH. Adaption in Natural and Artificial Systems. Quarterly Review of Biology. 1975;6(2):126–137.
31. Jakabosky J, Anglin P, Cook MR, et al. Non-linear FM waveform design using Marginal Fisher's Information within the CPM framework. In: 2011 IEEE RadarCon (RADAR); 2011. p. 513–518.
32. Blunt SD, Jakabosky J, Cook M, et al. Polyphase-coded FM (PCFM) radar waveforms, part II: optimization. IEEE Transactions on Aerospace and Electronic Systems. 2014;50(3):2230–2241.
33. Conn A, Gould N, Toint P. A globally convergent Lagrangian barrier algorithm for optimization with general inequality constraints and simple bounds. Mathematics of Computation of the American Mathematical Society. 1997;66(217):261–288.
34. Cumming IG, Wong FHc. Digital processing of synthetic aperture radar data: algorithms and implementation. 2005;.
35. Villano M, Krieger G, Moreira A. New Insights Into Ambiguities in Quad-Pol SAR. IEEE Transactions on Geoscience and Remote Sensing. 2017 ;55(6):3287–3308.

Chapter 5
Azimuth Phase Code for Ambiguity Suppression

The azimuth phase coding (APC) technique has been widely recognized in conventional synthetic aperture radar (SAR) systems, not only for its extremely low implementation complexity but also for being highly effective in suppressing both point and distributed ambiguities. In recent years, APC has also been investigated as an expansion to multichannel SAR systems. However, when multichannel SAR adopts digital beamforming (DBF) on receive, the desirable properties of APC can no longer be fully acquired, and only a minor improvement in range-ambiguity suppression can be achieved. In this section, we first provide a more comprehensive analysis of APC applied to multichannel SAR under a uniform pulse repetition frequency (PRF). We then briefly discuss APC/multichannel SAR configurations with nonuniform spatial sampling and introduce an enhanced reconstruction method based on a quadratically constrained optimization formulation which significantly increases the achievable APC gain compared with existing multichannel reconstruction algorithms, enabling the minimization of range ambiguity under a prescribed azimuth-ambiguity constraint. Notably, for certain specific PRFs, the above-mentioned method can eliminate odd-order range ambiguities. In the end, simulation results are presented to demonstrate the benefits and effectiveness of the proposed approach.

5.1 Introduction

Spaceborne synthetic aperture radar (SAR) has drawn growing interest from researchers worldwide, for it can operate under all-day and all-weather conditions [1–3]. Nevertheless, in conventional SAR systems, there are some difficulties in achieving both high spatial resolution and wide swath coverage due to ambiguity constraints [4, 5]. Therefore, our future focus will be on ambiguity suppression and the development of next-generation high-resolution wide-swath (HRWS) SAR systems [6–10].

R. Wang and W. Wang, *Synthetic Aperture Radar (SAR) Waveform Design and Signal Processing*, https://doi.org/10.1007/978-981-95-7549-7_5

Range ambiguities arise when echoes from the currently transmitted pulse are received only after a subsequent pulse has already been emitted [11, 12]. As a result, an excessively high pulse-repetition frequency (PRF) leads to ambiguous range information. To solve this issue, azimuth multichannel SAR employs spatial sampling through multiple receive channels in place of temporal oversampling, thereby reducing the required system PRF [13–19]. This substitution relaxes system design constraints, substantially suppresses range ambiguities, improves overall SAR performance, and alleviates the inherent tradeoff between swath width and spatial resolution. Therefore, azimuth multi-channel SAR provides an effective means to solve the contradiction between a high spatial resolution and a wide swath width.

The azimuth phase coding (APC) method introduced by Dall and Kusk [20] shifts the Doppler spectrum of range ambiguity by using the residual modulation phase produced through phase modulation of the transmitted signal and phase demodulation of the received echoes. Designed for conventional SAR systems with a single transmit and receive channel, this technique offers a simple implementation and can be applied uniformly to both point and distributed ambiguities, making it an effective approach for suppressing range ambiguities.

As noted earlier, the APC technique proposed by Dall and Kusk is originally designed for conventional SAR systems. Building on its applicability in such systems, Bordoni et al. [21] analyzed how APC behaves and what level of performance it can achieve when applied to a uniform multichannel SAR system employing DBF on receive. Unlike the single-channel case, the desirable properties of APC are no longer fully preserved in the multichannel configuration. Instead, APC in multichannel SAR exhibits several additional characteristics: (i) the APC gain remains above 0 dB; (ii) for a fixed number of channels, the APC gain grows with the oversampling factor; (iii) the APC gain decreases as the number of channels increases; and (iv) the sensitivity of the APC gain to the oversampling factor diminishes with more channels. Although the APC gain stays still greater than 0 dB, it reaches only about 1.5 dB at best in DBF-based multichannel SAR systems operating with uniform PRFs. Moreover, achieving uniform spatial sampling is typically impractical in multichannel implementations. For this reason, the performances of nonuniform spatial sampling should also be analyzed in APC/multichannel SAR systems.

Extending the ideas in [20, 21], this paper investigates the behaviors and attainable performance of APC in multichannel SAR systems using DBF on receive under nonuniform spatial sampling. Furthermore, due to the APC operation, the azimuth spectrum of the range ambiguity no longer aligns with that of the desired signal. This enables the formulation of a new reconstruction strategy capable of simultaneously recovering the desired signal and suppressing the range ambiguity.

The structure of this paper is as follows. Section 5.2 presents the signal model for multichannel SAR using the APC technique. Section 5.3 provides a more detailed analysis of APC/multichannel SAR with uniform spatial sampling. The discussion is then extended to nonuniform spatial sampling, where an enhanced reconstruction algorithm is introduced based on a quadratically constrained optimization framework. This method enables effective minimization of range ambiguity under a specified azimuth-ambiguity constraint and, for certain PRF choices, can cancel the range

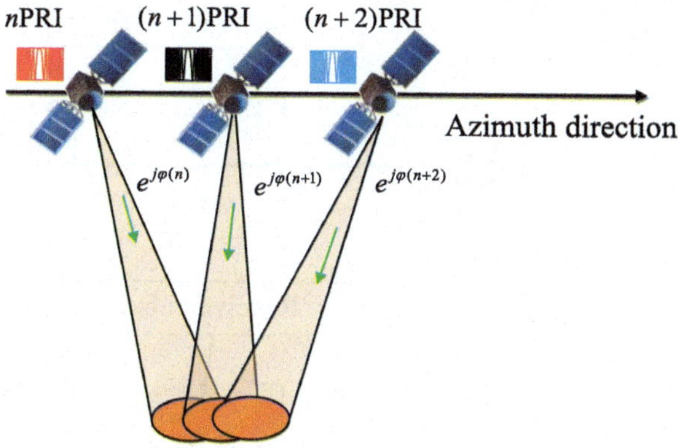

Fig. 5.1 Configuration for conventional SAR systems with azimuth phase coding

ambiguity entirely. Section 5.4 offers numerical analysis to assess how system performance varies with PRF, followed by a distributed-scene simulation demonstrating the effectiveness and advantages of the proposed reconstruction approach. Finally, Sect. 5.5 concludes the paper.

5.2 Azimuth Phase Code Method

5.2.1 APC Modulation and Demodulation

Phase pulse coding for SAR, also known as azimuth pulse coding, was firstly proposed by Dall and Kusk [20]. In general, APC refers to a method used to suppress range ambiguities in traditional SAR systems(one transmit, one receive channel), as illustrated in Fig. 5.1. The approach primarily involves three key processing steps:

(i) azimuth (i.e., pulse to pulse) phase modulation on transmission.
(ii) azimuth phase demodulation for each received pulse.
(iii) azimuth filtering over the processing bandwidth.

In the first step, a phase factor $e^{j\varphi(n)}$ is modulated to Mth transmit pulse by multiplying this phase factor with the transmit pulse, which can be seen in the up of Fig. 5.2. For spaceborne SAR systems, one may note that the round-trip delay time greatly exceeds the PRI, so the nth transmit pulse will be received as pulse number $n + m$.

In the second step, the demodulation process fully removes the modulation phase applied to the transmitted pulses for the strip of interest. This is achieved using a demodulation phase term $e^{-j\varphi(n-m)}$, as illustrated at the bottom of Fig. 5.2. Range ambiguities occur because echoes from preceding and succeeding pulse repetitions can return to the radar at the same time as the desired echo from the target strip. Consequently, a range ambiguity of order k, which corresponds to an extra two-

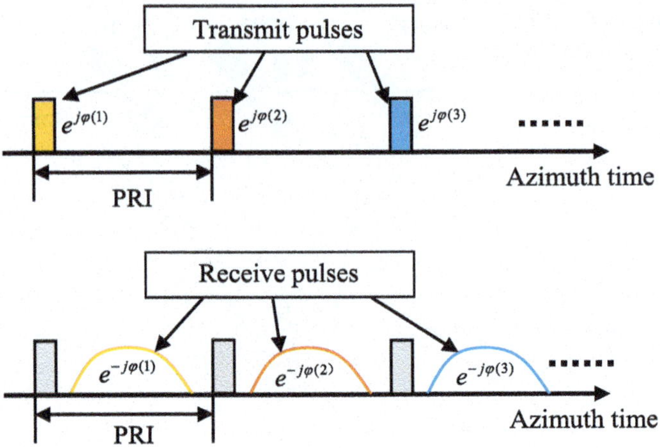

Fig. 5.2 Illustration of azimuth phase modulation on transmission (up) and demodulation on reception (bottom)

way propagation delay of k times the PRI, retains a residual modulation phase after undergoing the modulation and demodulation operations.

Further, the spectra of the range ambiguities are shifted in the frequency domain due to the residual modulation phase, as they no more superimpose the signal of interest, can be filtered out over the processing bandwidth in the third step.

In detail, the modulation of APC technique is given by

$$\varphi_{\mathrm{mod}}(n) = -\frac{\pi}{M_a} n^2 \tag{5.1}$$

where M denotes the transmit pulse number and M_a is a positive integer, denoted as APC shift factor. Then, the APC demodulation phase can be accordingly obtained, as

$$\varphi_{\mathrm{dem}}(n) = -\varphi_{\mathrm{mod}}(n-m) = \frac{\pi}{M_a}(n-m)^2 \tag{5.2}$$

As a consequence, after the above APC modulation and demodulation, the azimuth signal is affected by an APC residue phase, given by

$$\begin{aligned}\varphi_{\mathrm{res}}(n, k, M_a) &= \frac{\pi}{M_a}(n-m+k)^2 - \frac{\pi}{M_a}(n-m)^2 \\ &= \frac{2\pi}{M_a} kn + \frac{\pi}{M_a}(k^2 - 2km)\end{aligned} \tag{5.3}$$

where k represents the order of range ambiguity ($k = 0$ for the signal of interest, also named by the useful signal). One may note that, the last term of (5.3) is a constant independence on the pulse number M and can be neglected, and the first term will cause a shift to the Doppler spectrum of the kth range ambiguity.

Further, let $s(\eta)$ be the continuous azimuth signal of SAR systems, then the sampled azimuth signals can be expressed as

$$\tilde{s}(\eta) = \sigma \sum_{n} s(\eta)\delta(\eta - n\text{PRI}) \tag{5.4}$$

where $\delta(\eta)$ is the Dirac delta function, σ represents the scatter reflectivity and PRI is the repetition frequency interval. After the APC modulation and demodulation, the kth range ambiguity signal with the APC residue phase is given by

$$\tilde{s}_{\text{apc}}^{(k)}(\eta) = \sigma_k \sum_{n} s(\eta)\delta(\eta - n\text{PRI}) \exp\left(j\frac{2\pi}{M_a}kn\right) \tag{5.5}$$

where σ_k represents the scatter reflectivity of kth range ambiguity. It is worth noting that the above equation can be rewritten as

$$\tilde{s}_{\text{apc}}^{(k)}(\eta) = \sigma_k \sum_{n} s(\eta) \exp\left(j\frac{2\pi k}{M_a\text{PRI}}\eta\right) \delta(\eta - n\text{PRI}) \tag{5.6}$$

Thus, the azimuth spectrum of kth range ambiguity after the application of APC is given by

$$\tilde{S}_{\text{apc}}^{(k)}(f_\eta) = \sigma_k \sum_{m} S(f_\eta - mf_p - \Delta f),\ \Delta f = kf_p/M_a \tag{5.7}$$

where $S(f_\eta)$ is the Fourier transformation of $s(\eta)$. Therefore, the spectra of the range ambiguities are shifted in the frequency domain due to the APC residue phase. Moreover, the spectra have a periodic behavior versus the PRF, and the actual APC Doppler shift is given by

$$\Delta f = \text{mod}[kf_p/M_a]_{f_p} \tag{5.8}$$

where $\text{mod}[\cdot]_{f_p}$ denotes the modulus operator. Further, the APC Doppler shift belongs to the interval $[0, f_p]$. Figure 5.3 illustrates the basic effect of the APC residue phase in the frequency domain ($M_a = 3$): the rang ambiguous signal of order $k = 1$ and $k = 2$ suffer from APC Doppler shifts, $\Delta f = f_p/3, 2f_p/3$, respectively, while the spectrum of the useful signal is unchanged. As a result, the part of the range ambiguous power located outside the processed bandwidth B_d can be filtered out by the third APC step. Moreover, one can also note that the ambiguity components can be largely filtered out when $\Delta f = f_p/2$.

Generally, the range ambiguity of first order provides the dominant contribute to the SAR image quality degradation, and only $M_a = 2$ allows for the best suppression of the range ambiguity of first order (in this case, $\Delta f = f_p/2$). Therefore, it is particularly significant to analyze the APC performance for $M_a = 2$ and maximize ambiguity suppression.

Fig. 5.3 Simplified representation of the APC effect in frequency domain ($M_a = 3$): useful signal (solid) and range ambiguities of order $k = 1, 2$ (dash-dotted, dashed). The shadowed ambiguity components can be filtered out

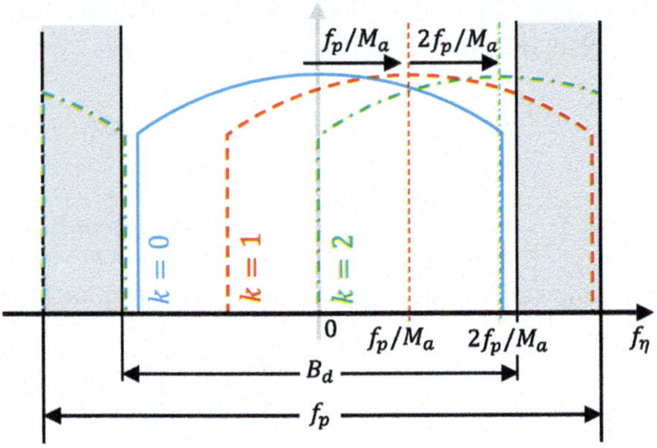

5.2.2 *APC Gain for Conventional SAR System*

To describe how applying the APC technique influences the resulting SAR image, the APC gain is introduced as a metric that quantifies the level of range-ambiguity suppression achieved and the corresponding enhancement in image quality. This gain is evaluated from the azimuth spectra of the range-ambiguous components and represents the decrease in their power caused by the APC processing. In particular, it is given by the power ratio of range ambiguous signals before and after the application of the APC technique.

In detail, the azimuth spectrum of kth range ambiguity (without application of APC) is given by

$$\tilde{S}^{(k)}(f_\eta) = \sigma_k \sum_m S(f_\eta - m f_p) \tag{5.9}$$

Then, the power of all range ambiguous signals over the processing bandwidth can be expressed as

$$\begin{aligned} P^{\mathrm{am}} &= \sum_{k \neq 0} \int_{-B_d/2}^{B_d/2} E\left[|\tilde{S}^{(k)}(f_\eta)|^2 \right] df_\eta \\ &= \sum_{k \neq 0} \sigma_k^2 \sum_{m=-\infty}^{\infty} \int_{-B_d/2}^{B_d/2} |S(f_\eta - m f_p)|^2 \, df_\eta \end{aligned} \tag{5.10}$$

Similarly, the power of all range ambiguous signals with the application of APC is given by

Table 5.1 Parameters of the planar SAR systems used for APC performance analysis

Parameters	Value
Orbit height	520 km
Tx antenna length	3 m
Rx antenna length	3 m
Carrier frequency	9.6 GHz
Processed bandwidth	2104–4208 Hz
Oversampling factor	1.2 ~ 2.4
PRF	5050 Hz
Platform velocity	7600 m/s

$$
\begin{aligned}
P_{\text{apc}}^{\text{am}} &= \sum_{k\neq 0} \int_{-B_d/2}^{B_d/2} E\left[|\tilde{S}_{\text{apc}}^{(k)}(f_\eta)|^2\right] df_\eta \\
&= \sum_{k\neq 0} \sigma_k^2 \sum_{m=-\infty}^{\infty} \int_{-B_d/2}^{B_d/2} |S(f_\eta - mf_p - \Delta f)|^2 df_\eta
\end{aligned} \tag{5.11}
$$

After that, the APC gain for conventional SAR systems can be accordingly defined as

$$
G_{\text{apc}} = \frac{P^{\text{am}}}{P_{\text{apc}}^{\text{am}}} \tag{5.12}
$$

In particular, for $M_a = 2$, the APC gain can be expressed as

$$
\begin{aligned}
G_{\text{apc}} &= \frac{\sum_{k\neq 0} \sigma_k^2 \sum_{m=-\infty}^{\infty} \int_{-B_d/2}^{B_d/2} |S(f_\eta - mf_p)|^2 df_\eta}{P_{odd} + P_{even}} \\
P_{odd} &= \sum_{k=1}^{N_{odd}} \sigma_k^2 \sum_{m=-\infty}^{\infty} \int_{-B_d/2}^{B_d/2} |S(f_\eta - \frac{2m+1}{2} f_p)|^2 df_\eta \\
P_{even} &= \sum_{k=1}^{N_{even}} \sigma_k^2 \sum_{m=-\infty}^{\infty} \int_{-B_d/2}^{B_d/2} |S(f_\eta - mf_p)|^2 df_\eta
\end{aligned} \tag{5.13}
$$

where N_{odd} and N_{even} are the numbers of odd and even range ambiguities, respectively, and where it has been assumed for simplicity that the azimuth spectra are equal for all range ambiguities.

For the performance evaluation, a conventional SAR system equipped with a planar antenna is defined. The parameters used to compute the APC gain are provided in Table 5.1. The transmit and receive antennas have the same azimuth length, and thus their azimuth patterns are identical, as illustrated in Fig. 5.4.

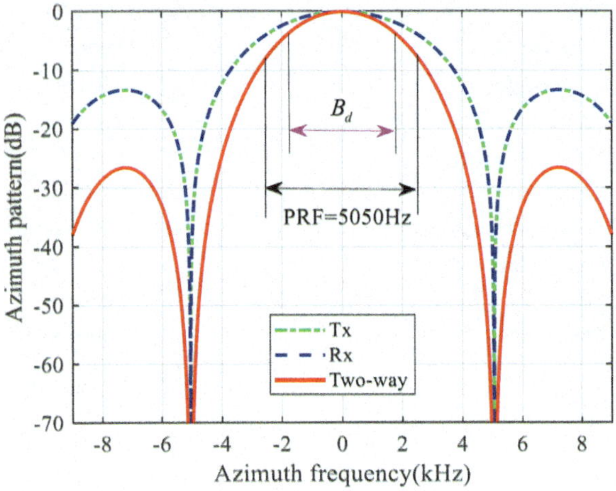

Fig. 5.4 Azimuth pattens (solid: two-way; dash-dot: Tx; dashed: Rx) for the conventional SAR systems. As an example, the processed bandwidth $B_d = 3544$ Hz and PRF = 5050 Hz

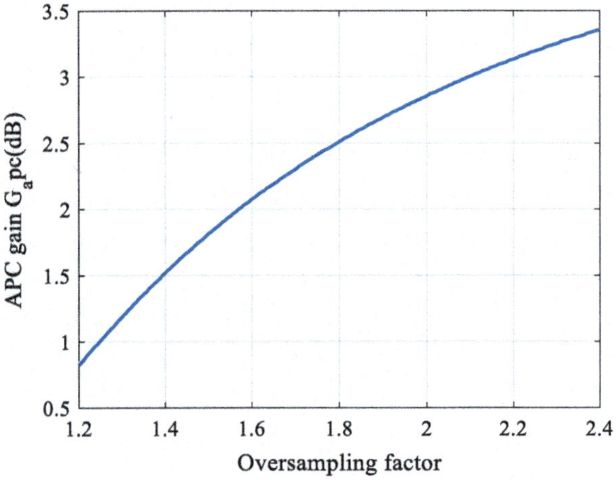

Fig. 5.5 APC gain versus the oversampling factor for the conventional SAR systems

The oversampling factor is defined as the ratio of the PRF to the processed bandwidth:

$$\gamma = \frac{\text{PRF}}{B_d} \tag{5.14}$$

In particular, it is assumed that the oversampling factor ranges from 1.2 to 2.4. Accordingly, the processing bandwidth 2104 Hz$\leq B_d \leq$ 4208 Hz has been considered.

Figure 5.5 shows the APC gain, G_{apc}, versus the oversampling factor, obtained for the SAR system described in Table 5.1. In this case, the APC shift-factor is assumed equal to two $M_a = 2$ and only the range ambiguity of first order is considered. One can observe from Fig. 5.5 that the APC gain ranges from 0.83 dB to 3.36 dB, and as expected, a larger G_{apc} is associated with a larger γ. In fact, the power spectral density

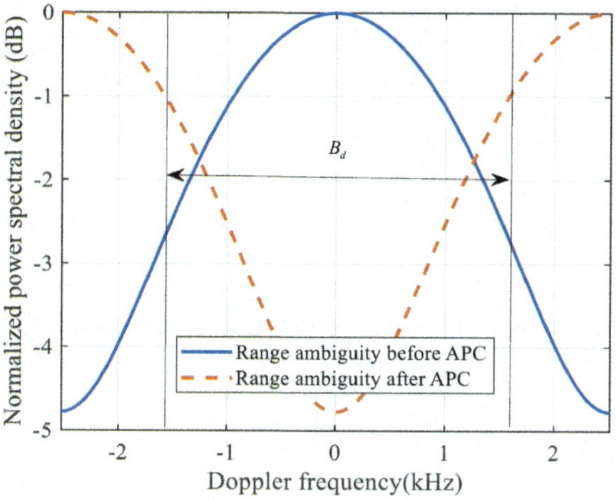

Fig. 5.6 The obtained PSD over the PRF interval of the first range ambiguity before and after APC

(PSD) of first range ambiguity is shifted by $\Delta f = 2525$ Hz after APC (dashed curve), and thus when the oversampling is higher, a larger part of PSD of first ambiguity falls outside the processing bandwidth after APC. In contrast, a large part of PSD of first ambiguity falls inside the processing bandwidth without the application of APC. This is illustrated in Fig. 5.6, which shows the obtained PSD over the PRF interval of the first range ambiguity before and after APC.

5.3 The Application of APC in Multichannel SAR

5.3.1 *Signal Model with APC in Multichannel SAR*

Without loss of generality, this article focuses solely on the APC shift factor $M = 2$ [20]. For this choice, the corresponding APC modulation phase can be readily derived by

$$\varphi_{mod}(m) = -\frac{\pi}{2}m^2 \tag{5.15}$$

where m denotes the transmit pulse number. Obviously, $\varphi(m)$ assumes alternatively the values 0 and $-\frac{\pi}{2}$. After modulation and demodulation, the azimuth signal is affected by the APC residual phase

$$\varphi_{res}(p, q) = -\pi qp, \quad q = 0, \pm 1, \pm 2, \ldots \tag{5.16}$$

where p is the receive pulse number and q is the order of range ambiguity ($q = 0$ for the signal of interest).

Consequently, when APC is applied in a multichannel SAR system, the desired signal remains unaffected by the residual APC phase, whereas the odd-order range-ambiguous components exhibit a signal model that differs from that of the signal of interest.

5.3.1.1 Signal Model with APC

The APC technique was originally proposed by Dall and Kusk [20] to suppress range ambiguities in SAR systems. It was later extended to multichannel SAR configurations in [21]. Consider a multichannel SAR system equipped with M uniformly spaced antennas along the azimuth (along-track) direction, with an inter-element spacing of d_a. The received signal for the jth channel ($j = 1, 2, \ldots, M$) can be expressed, following the formulation in [13], in terms of the impulse response of a reference channel, which can be approximated as

$$s_j^0(\eta) \approx s^0\left(\eta - \frac{d_j}{2V_r}\right) \tag{5.17}$$

where $s^0(\eta)$ represents the impulse response of the reference channel (taken here, without loss of generality, as the first channel), $d_j = (j-1)d_a$ is the azimuth displacement between the reference channel and the jth channel and V_r denotes the platform velocity.

Another perspective for describing azimuth aliasing–one that is also useful for deriving the ambiguity-to-signal ratio (AASR) in multichannel SAR systems after signal reconstruction–is to represent the sampled azimuth signa $\tilde{s}_j^0(\eta)$ for the jth channel as the product of its continuous azimuth-domain signal $s_j^0(\eta)$ and a Dirac comb. This formulation can be written as

$$\tilde{s}_j^0(\eta) = s_j^0(\eta) \sum_k \delta(\eta - k\mathrm{PRI}) \tag{5.18}$$

where PRI is the time between two transmitted pulses. According to the property of Fourier series [22],

$$\sum_k \delta(\eta - k\mathrm{PRI}) = \frac{1}{\mathrm{PRI}} \sum_k \exp(j2\pi k f_p \eta) \tag{5.19}$$

where $f_p = 1/\mathrm{PRI}$ is the pulse repetition frequency (PRF). Then, the spectrum of the sampled azimuth signal $\tilde{s}_j^0(\eta)$ is given by

$$\tilde{S}_j^0(f) = \frac{1}{\text{PRI}} \sum_k S_j^0(f_\eta - kf_p)$$
$$= \frac{1}{\text{PRI}} \sum_k S^0(f_\eta - kf_p) H_j(f_\eta - kf_p) \quad (5.20)$$

where f_η denotes the Doppler frequency and the influence of the multichannel system is described by $H_j(f_\eta)$,

$$H_j(f_\eta) = \exp\left(-j\frac{\pi d_j}{V_r} f_\eta\right) \quad (5.21)$$

In the absence of APC, the range-ambiguous echoes received in a multichannel SAR system share the same azimuth-spectrum characteristics as the echoes originating from the desired swath. When APC is applied, however, the azimuth spectrum of the desired signal remains unchanged, whereas the spectrum of the range-ambiguous components is modified by the APC operation. Let $s^r(\eta)$ denote the range-ambiguous signal in the case without APC; under APC, the corresponding range-ambiguous signal received by the jth channel can then be approximated as

$$s_j^r(\eta) \approx s^r\left(\eta - \frac{d_j}{2V_r}\right) \quad (5.22)$$

With the use of the APC technique ($M = 2$), the sampled azimuth signal $\tilde{s}_j^r(\eta)$ from the range ambiguity can be expressed as

$$\tilde{s}_j^r(\eta) = s_j^r(\eta) \sum_k \delta(\eta - 2k\text{PRI}) - s_j^r(\eta) \sum_k \delta(\eta - (2k+1)\text{PRI}) \quad (5.23)$$

According to Eq. (5.19), the corresponding spectrum $\tilde{S}_j^r(f_\eta)$ of signal $\tilde{s}_j^r(\eta)$ can be obtained as follows:

$$\tilde{S}_j^r(f_\eta) = \frac{1}{\text{PRI}} \sum_k S^r(f_\eta - kf_p - f_p/2) H_j(f_\eta - kf_p - f_p/2) \quad (5.24)$$

where $S^r(f_\eta)$ is the spectrum of the signal $s^r(\eta)$.

According to equations (5.20) and (5.24), the azimuth spectrum of the signal of interest and the range-ambiguous signal with noise $\boldsymbol{n}(f_\eta)$ in vector format can be reformulated as

$$\tilde{\boldsymbol{S}}^0(f_\eta) = \boldsymbol{A}(f_\eta)\boldsymbol{S}^0(f_\eta) + \boldsymbol{n}(f_\eta) \quad (5.25)$$

and

$$\tilde{\boldsymbol{S}}^r(f_\eta) = \boldsymbol{C}_r\boldsymbol{A}(f_\eta)\boldsymbol{S}^r(f_\eta) + \boldsymbol{n}(f_\eta) \quad (5.26)$$

where the signal of interest in vector format is given by

$$\boldsymbol{S}^0(f_\eta) = [\ldots, S^0(f_\eta), \ldots, S^0(f_\eta - kf_p), \ldots]^T \tag{5.27}$$

and the range-ambiguous signal is given by

$$\boldsymbol{S}^r(f_\eta) = [\ldots, S^r(f_\eta - f_p/2), \ldots, S^r(f_\eta - kf_p - f_p/2), \ldots]^T \tag{5.28}$$

where $(\cdot)^T$ denotes the vector transpose operation. The matrix $\boldsymbol{A}(f_\eta)$ represents the channel matrix given by

$$\boldsymbol{A}(f_\eta) = [\ldots, \boldsymbol{a}_0(f_\eta), \ldots, \boldsymbol{a}_k(f_\eta), \ldots] \tag{5.29}$$

$$\boldsymbol{a}_k(f_\eta) = \boldsymbol{a}(f_\eta - kf_p) = \begin{bmatrix} e^{-j\pi(f_\eta - k\cdot f_p)d_1/V_r} \\ \vdots \\ e^{-j\pi(f_\eta - k\cdot f_p)d_j/V_r} \\ \vdots \\ e^{-j\pi(f_\eta - k\cdot f_p)d_M/V_r} \end{bmatrix} \tag{5.30}$$

The matrix $\boldsymbol{C}_r$ is induced by the APC technique and only depends on the PRF.

$$\boldsymbol{C}_r = \mathrm{diag}\big([e^{j\pi f_p d_1/2V_r}, \ldots, e^{j\pi f_p d_j/2V_r}, \ldots, e^{j\pi f_p d_M/2V_r}]\big) \tag{5.31}$$

where $\mathrm{diag}(\boldsymbol{x})$ denotes a diagonal matrix whose diagonal elements are the components of a vector $\boldsymbol{x}$. Then, the signal-plus-noise covariance matrix of interest corresponding to the multichannel output in Doppler bin f_η is given by

$$\begin{aligned} \boldsymbol{R}_0 &= E(\tilde{\boldsymbol{S}}^0(f_\eta)\tilde{\boldsymbol{S}^0}^H(f_\eta)) \\ &= \boldsymbol{A}(f_\eta)E\big(\boldsymbol{S}^0(f_\eta)\boldsymbol{S}^{0H}(f_\eta)\big)\boldsymbol{A}^H(f_\eta)\big) + \sigma_n^2\boldsymbol{I}_N \end{aligned} \tag{5.32}$$

Similarly, the range-ambiguity signal-plus-noise covariance matrix is given by

$$\begin{aligned} \boldsymbol{R}_r &= E(\tilde{\boldsymbol{S}}^r(f_\eta)\tilde{\boldsymbol{S}^r}^H(f_\eta)) \\ &= \boldsymbol{C}_r\boldsymbol{A}(f_\eta)E\big(\boldsymbol{S}^r(f_\eta)\boldsymbol{S}^{rH}(f_\eta))\boldsymbol{A}^H(f_\eta)\big)\boldsymbol{C}_r^H + \sigma_n^2\boldsymbol{I}_N \end{aligned} \tag{5.33}$$

where $E(\cdot)$ denotes the statistical expectation, $(\cdot)^H$ denotes the matrix conjugate transpose operator, σ_n^2 is the noise power and $\boldsymbol{I}_N$ is the identity matrix of order M. In the case of SAR imaging, the covariance matrices $\boldsymbol{R}_0$ and $\boldsymbol{R}_r$ can be reformulated as [17, 18]

$$\boldsymbol{R}_0 = \sigma_s^2 \boldsymbol{A}(f_\eta)\boldsymbol{\Lambda}_0\boldsymbol{A}^H(f_\eta) + \sigma_n^2\boldsymbol{I}_N \tag{5.34}$$

and

$$\boldsymbol{R}_r = \sigma_r^2 \boldsymbol{C}_r\boldsymbol{A}(f_\eta)\boldsymbol{\Lambda}_r\boldsymbol{A}^H(f_\eta)\boldsymbol{C}_r^H + \sigma_n^2\boldsymbol{I}_N \tag{5.35}$$

where σ_t^2 and σ_r^2 represent the signal power of interest and the range-ambiguous signal power, respectively, and $\boldsymbol{\Lambda}_0$ and $\boldsymbol{\Lambda}_r$ are given by

$$\boldsymbol{\Lambda}_0 = \text{diag}[\ldots, |S^0(f_\eta - kf_p - f_p/2)|^2, \ldots] \tag{5.36}$$

and

$$\boldsymbol{\Lambda}_r = \text{diag}[\ldots, |S^r(f_\eta - kf_p - f_p/2)|^2, \ldots] \tag{5.37}$$

5.3.1.2 Sampling Frequency (PRF)

A few considerations regarding the PRF and its relationship to the antenna spacing d are in order. Reference [17] defines a specific set of PRFs given by

$$\text{PRF}^{\text{SPEC}} = \frac{2V_r}{d_a}\frac{1}{K}, \; K \in \{1, \ldots, M-1\} \tag{5.38}$$

When using PRFs from this set, the channel vector $\boldsymbol{a}_k$ is periodic with period K

$$\boldsymbol{a}_k = \boldsymbol{a}_{k+K} \tag{5.39}$$

As a result, the number of degrees of freedom (DOFs) K available to reconstruct the signal is lower than available Rx channels M. When K equals the number of channels M, the uniform PRF can be accordingly defined as

$$\text{PRF}_{\text{uni}} = \frac{2V_r}{d_a}\frac{1}{M} \tag{5.40}$$

Then, nonuniform PRFs are between the uniform PRF PRF_{uni} and specific PRFs PRF^{SPEC}.

5.3.2 *Ambiguity Suppression Based on APC in a Multichannel SAR System*

Recently, a multichannel SAR system employing the APC technique with uniform spatial sampling was investigated in [21]. However, the extension of APC to nonuniform multichannel SAR configurations has not yet been addressed. In this section, we first present a more detailed analysis of APC in multichannel SAR systems with uniform spatial sampling. Subsequently, we discuss APC in multichannel systems with nonuniform spatial sampling and introduce a novel reconstruction algorithm for ambiguity suppression.

5.3.2.1 Uniform PRF

In fact, uniform spatial sampling in a multichannel SAR system is equivalent to uniform temporal sampling in a single-channel SAR system. As previously noted, the uniform PRF is defined by (5.40). When $f_p = \mathrm{PRF_{uni}}$, the equivalent time shift for the jth channel relative to the reference channel is given by

$$\frac{d_j}{2V_r} = \frac{(j-1)}{M\mathrm{PRF_{uni}}} = \frac{(j-1)}{Mf_p} = \frac{(j-1)\mathrm{PRI}}{M} \tag{5.41}$$

To evaluate the behavior of APC with respect to the number of receivers, systems with 3, 2, and 1 receive channels were considered (see systems #3 and #2, #1 in Table 5.2. Their parameters [21] used for computing APC performance are summarized in Table 5.2, and the corresponding space-time sampling patterns are illustrated in Fig. 5.7.

Let PRI denote the pulse-repetition interval (PRI) of system #1; then, the PRIs of systems #2 and system #3 are 2PRI and 3PRI, respectively, as shown in Fig. 5.7. It can be observed that the desired signal maintains the same phase, indicated by '+', whereas the range-ambiguous signal exhibits alternating phases, denoted by '+' and '−'.

Bordoni et al. [21] concluded that (i) for a given number of channels M, the APC gain increases with the oversampling factor, and (ii) the APC gain decreases as the number of channels M increases. In the following, a more detailed mathematical analysis of this APC gain is provided.

Table 5.2 Performances of the SAR systems used for performance analysis

Parameter	System #1	System #2	System #3
Orbit height		520 km	
Carrier frequency		9.6 GHz	
Subaperture length		3 m	
Total receiver length	3 m	6 m	9 m
Number of receivers	1	2	3
PRF	5070 Hz	2535 Hz	1690 Hz
Effective PRF		5070 Hz	

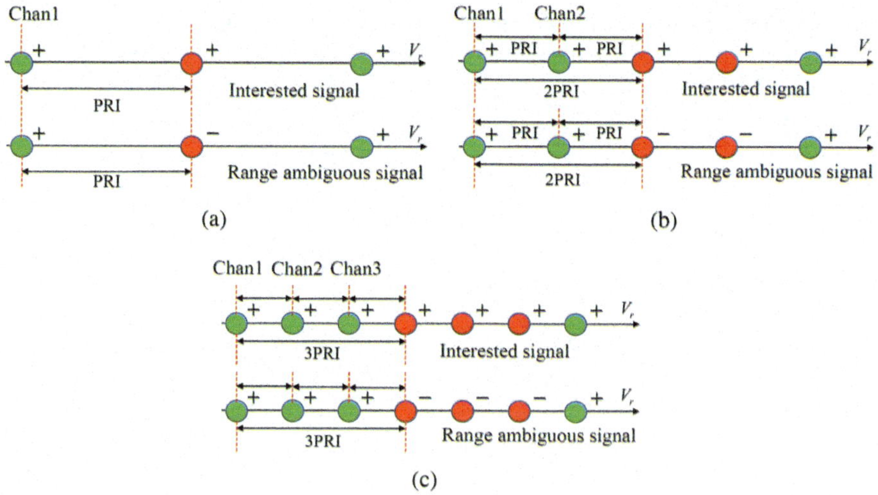

Fig. 5.7 Space-time sampling patterns. The green and red circles represent different pulses. The '+' denotes same phase modulation, while the '−' represents opposite phase modulation. **a** System #1. **b** System #2. **c** System #3

Under a uniform PRF, the spatial sampling of the range-ambiguous signal in a multichannel SAR system is equivalent to temporal sampling in a single-channel SAR system. Consequently, the sampled range-ambiguous signal of system #1 after APC modulation and demodulation can be expressed as

$$\tilde{s}^{r1}(\eta) = s^r(\eta)\sum_k \delta(\eta - 2k\text{PRI}) - s^r(\eta)\sum_k \delta(\eta - 2k\text{PRI} - \text{PRI}) \tag{5.42}$$

Then, the corresponding spectrum is given by

$$\tilde{S}^{r1}(f_\eta) = \frac{1}{\text{PRI}}\sum_k S^r(f_\eta - kf_p - f_p/2) \tag{5.43}$$

Thus, the spectrum of the range-ambiguous signal suffers from a frequency shift $f_p/2$ relative to that of the signal of interest, and the period of this spectrum is f_p. For system #2, the sampled range-ambiguous signal is written as

$$\begin{aligned}\tilde{s}^{r2}(\eta) = {} & s^r(\eta)\sum_k \delta(\eta - 4k\text{PRI}) + s^r(t)\sum_k \delta(\eta - (4k+1)\text{PRI}) \\ & - s^r(\eta)\sum_k \delta(\eta - (4k+2)\text{PRI}) \\ & - s^r(\eta)\sum_k \delta(\eta - (4k+3)\text{PRI})\end{aligned} \tag{5.44}$$

Table 5.3 Spectrum characteristic of these three systems with $f_p = 5070$ Hz

Parameter	Period	Frequency shift	Effective PRF	PRI
System #1	f_p	$f_p/2$	5070 Hz	T
System #2	$f_p/2$	$f_p/4$	5070 Hz	2T
System #3	$f_p/3$	$f_p/6$	5070 Hz	3T

The corresponding spectrum is given by

$$\tilde{S}^{r2}(f_\eta) = \frac{1}{2\text{PRI}} \sum_k S^r(f_\eta - kf_p/2 - f_p/4) + \frac{j}{2\text{PRI}} \sum_k (-1)^k S^r(f_\eta - kf_p/2 - f_p/4) \tag{5.45}$$

Thus, the period of the range-ambiguous signal spectrum in system #2 is $f_p/2$. Similarly, one can obtain the spectrum of the sampled range-ambiguous signal of system #3 as follows:

$$\begin{aligned}\tilde{S}^{r3}(f) = &\frac{1}{3\text{PRI}} \sum_k S^r(f_\eta - kf_p/3 - f_p/6) \\ &+ \frac{1}{3\text{PRI}} \sum_k e^{j\frac{k\pi}{3}} S^r(f_\eta - kf_p/3 - f_p/6) \\ &+ \frac{1}{3\text{PRI}} \sum_k e^{j\frac{2k\pi}{3}} S^r(f_\eta - kf_p/3 - f_p/6)\end{aligned} \tag{5.46}$$

The corresponding spectrum period of the range-ambiguous signal in system #3 is $f_p/3$. Table 5.3 summarizes the spectral characteristics of these three systems.

If these three systems have identical azimuth antenna patterns and the PRF exceeds the Doppler bandwidth, the range-ambiguous signals in systems #2 and system #3 will undergo significant azimuthal aliasing. The corresponding spectra of the three systems are illustrated in Fig. 5.8.

As indicated in Table 5.3, the spectral period of the range-ambiguous signal decreases as the number of channels M increases. Table 5.3 also shows that the frequency shift diminishes with more channels. Consequently, Fig. 5.8 reveals that the azimuth spectrum components of the range-ambiguous signal within the band $[-B_d/2, B_d/2]$ grow with increasing M. Furthermore, for a fixed number of channels, the APC gain rises as the Doppler bandwidth B_d decreases.

These observations indicate that the mathematical analysis of APC gain aligns well with the simulation results reported in [21]. In other words, when APC is applied to a multichannel SAR system with a uniform PRF, the APC gain–although positive–decreases as the number of channels increases. Additionally, simulations in [21] show

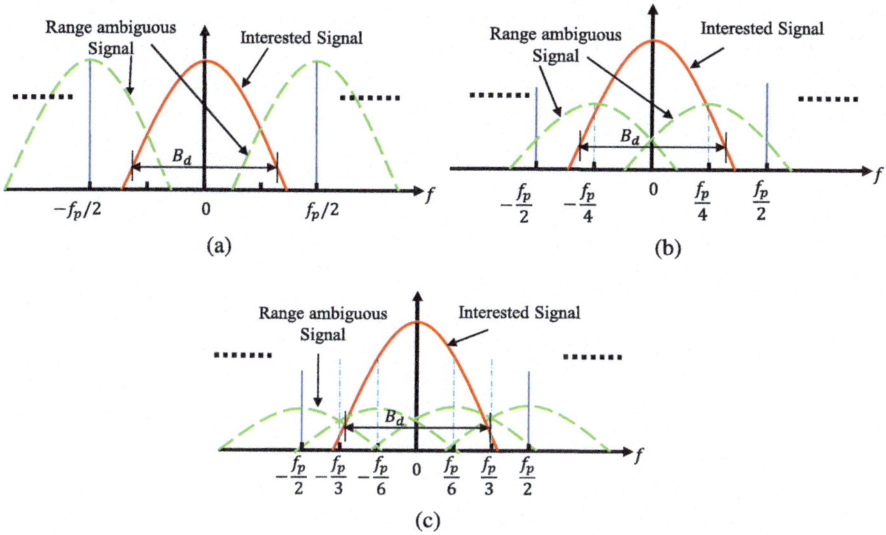

Fig. 5.8 The spectra of the three systems. The red and green lines represent the spectrum of the signal of interest and the range-ambiguous signal, respectively. **a** System #1. **b** System #2. **c** System #3

that the APC gain remains below 2 dB for a uniform PRF, which is insufficient to enhance SAR imaging performance. The following section investigates the APC gain in multichannel SAR systems under nonuniform PRFs.

5.3.2.2 Nonuniform PRFs

For nonuniform PRFs, spatial sampling in a multichannel SAR system no longer corresponds directly to uniform temporal sampling in a single-channel SAR system. In this context, Krieger [13] proposed a digital beamforming (DBF) reconstruction algorithm to recover the Doppler spectrum. Without loss of generality, the number of channels M is assumed to be odd. Let $\boldsymbol{\omega}(f_\eta - kf_p), k = (M-1)/2, \ldots, -(M-1)/2$ denote the set of DBF reconstruction filters; then, the kth reconstructed spectral component of the azimuth signal can then be expressed as

$$S_c(f_\eta - kf_p) = \boldsymbol{\omega}^H(f_\eta - kf_p)\big(\tilde{\boldsymbol{S}}^0(f_\eta) + \tilde{\boldsymbol{S}}^r(f_\eta)\big) \quad (5.47)$$

where the set of filters $\boldsymbol{\omega}(f_\eta - kf_p)$ can be obtained by computing the inverse of the $M \times M$ channel matrix $\boldsymbol{A}_0(f_\eta)$, which is a subset of matrix $\boldsymbol{A}(f_\eta)$. The $M \times 1$ filter $\boldsymbol{\omega}(f_\eta)$ is written by

$$\boldsymbol{\omega}(f_\eta - kf_p) = \boldsymbol{A}_0^{H^{-1}}(f_\eta)\boldsymbol{e}_k \quad (5.48)$$

where $\boldsymbol{e}_k = [0, 0, \ldots, 1, \ldots, 0]^T$ is an $M \times 1$ vector of zeros except for the kth element, which is equal to one.

As a consequence, the whole reconstructed Doppler spectrum can be obtained by adding an associated low-pass filter.

$$S_c(f) = \left[S_c\left(f_\eta - \frac{N-1}{2} f_p\right), \ldots, S_c\left(f_\eta + \frac{N-1}{2} f_p\right) \right] \cdot \text{rect}\left(\frac{f}{B_d}\right) \tag{5.49}$$

where rect(x) is a rectangular function defined as

$$\text{rect}(x) = \begin{cases} 1, & x \in [-0.5, 0.5] \\ 0, & \text{Others} \end{cases} \tag{5.50}$$

When APC is applied in a multichannel SAR system, the channel matrix of the desired signal remains unchanged, whereas the channel matrix of the range-ambiguous signal differs from $\boldsymbol{A}(f_\eta)$. As a result, the reconstructed spectrum of the range-ambiguous signal using DBF reconstruction filters is affected by the modified matrix $\boldsymbol{C}_r \boldsymbol{A}(f_\eta)$.

For performance analysis, the whole azimuth ambiguity power of the signal of interest can be defined as

$$\begin{aligned} P_{am}^0 &= \sum_{k=-\infty, \neq 0}^{\infty} \int_{-B_d/2}^{B_d/2} |\boldsymbol{\omega}^H(f_\eta) \boldsymbol{a}_k(f_\eta) S^0(f_\eta - k f_p)|^2 df_\eta \\ &= \sum_{k=-\infty}^{\infty} \int_{-B_d/2}^{B_d/2} \boldsymbol{\omega}^H(f_\eta) \boldsymbol{R}_0(f_\eta) \boldsymbol{\omega}(f_\eta) df_\eta \\ &\quad - \int_{-B_d/2}^{B_d/2} |S^0(f_\eta)|^2 df_\eta \end{aligned} \tag{5.51}$$

and the signal power of interest is written as

$$P_{\text{sig}}^0 = \int_{-B_d/2}^{B_d/2} |S^0(f_\eta)|^2 df_\eta \tag{5.52}$$

Then, the azimuth ambiguity-to-signal ratio (AASR) for the multichannel SAR system can be accordingly defined as

$$\text{AASR} = \frac{P_{am}^0}{P_{\text{sig}}^0} \tag{5.53}$$

IF APC is not employed, the whole range-ambiguous signal power can be defined as

$$P^r = \sum_{k=-\infty}^{\infty} \int_{-B_d/2}^{B_d/2} |\boldsymbol{\omega}^H(f_\eta)\boldsymbol{a}_k(f_\eta)S^r(f_\eta - kf_p)|^2 df_\eta \tag{5.54}$$

If APC is employed, the whole range-ambiguous signal power can be defined as

$$\begin{aligned} P^r_{\text{APC}} &= \sum_{k=-\infty}^{\infty} \int_{-B_d/2}^{B_d/2} |\boldsymbol{\omega}^H(f_\eta)\boldsymbol{C}_r\boldsymbol{a}_k(f_\eta) \cdot S^r(f_\eta - kf_p - f_p/2)|^2 df_\eta \\ &= \sum_{k=-\infty}^{\infty} \int_{-B_d/2}^{B_d/2} \boldsymbol{\omega}^H(f_\eta)\boldsymbol{C}_r^H\boldsymbol{R}_r(f_\eta)\boldsymbol{C}_r\boldsymbol{\omega}(f_\eta)df_\eta \end{aligned} \tag{5.55}$$

Then, the APC gain for a multichannel SAR system with nonuniform PRF can be accordingly defined as

$$G_{\text{APC}} = \frac{P^r}{P^r_{\text{APC}}} \tag{5.56}$$

When the PRF is uniformly spatially sampled, the DBF reconstruction filter essentially reduces to an equivalent temporal sampling in a single-channel SAR system. As a result, the APC gain achievable using the DBF reconstruction filter in a multichannel SAR system is expected to be lower than that of a single-channel SAR system under nonuniform spatial sampling. In other words, a multichannel SAR system combined with APC cannot suppress range ambiguities as effectively as a single-channel system. This limitation is confirmed in multichannel SAR systems operating with a uniform PRF.

To address this limitation, a novel reconstruction algorithm is proposed. Notably, when APC is applied in a multichannel SAR system, the channel function matrices of the desired signal and the range-ambiguous signal differ. This difference enables the design of improved filters that can suppress range-ambiguous signals without compromising azimuth ambiguity. The proposed reconstruction algorithm is described as follows.

First, we impose an ambiguity constraint on the power of the azimuth ambiguity

$$\boldsymbol{\omega}^H(f_\eta)\boldsymbol{R}_0(f_\eta)\boldsymbol{\omega}(f_\eta) <= \epsilon^2(f_\eta) \tag{5.57}$$

where $\epsilon^2(f_\eta)$ is a design constant that only varies with Doppler frequency f_η intended to relax the limitation of azimuth ambiguity. Although the APC is employed in a multichannel SAR system, the maximum suppression of range-ambiguous signal power cannot be obtained by the DBF reconstruction filter. For this reason, we establish an optimization model to minimize range-ambiguous signal power with a constraint shown in (5.57).

The optimization problem can be accordingly formulated as

$$\mathscr{P}:\begin{cases}\min\limits_{\boldsymbol{\omega}(f_\eta)} & \boldsymbol{\omega}^H(f_\eta)\boldsymbol{C}_r^H\boldsymbol{R}_r(f_\eta)\boldsymbol{C}_r\boldsymbol{\omega}(f_\eta)\\ \text{s.t.} & \boldsymbol{\omega}^H(f_\eta)\boldsymbol{a}(f_\eta)=1\\ & \boldsymbol{\omega}^H(f_\eta)\boldsymbol{R}_0(f_\eta)\boldsymbol{\omega}(f_\eta)\le\epsilon^2(f_\eta)\end{cases}\tag{5.58}$$

where the Doppler frequency f_η is within $[-B_d/2, B_d/2]$ and the constraint $\boldsymbol{\omega}^H\boldsymbol{a}=1$ is intended to ensure the Doppler spectrum of the signal of interest.

A solution to $\mathscr{P}$ can be found in [23, 24] using the method of Lagrange multipliers. The solution process is as follows.

For the sake of simplicity, the symbol f_η will be removed in the following discussion. Let $\mathscr{S}$ be the set defined by the constraints in (5.58). Obviously, matrices $\boldsymbol{C}_r^H\boldsymbol{R}_r\boldsymbol{C}_r$ and $\boldsymbol{R}_0$ are nonnegative definite matrices. Thus, let

$$\begin{aligned}f_1(\boldsymbol{\omega},\lambda,\mu)=\boldsymbol{C}_r^H\boldsymbol{R}_r\boldsymbol{C}_r+\lambda(\boldsymbol{\omega}^H\boldsymbol{R}_0\boldsymbol{\omega}-\epsilon^2)\\ +\mu(-\boldsymbol{\omega}^H\boldsymbol{a}-\boldsymbol{a}^H\boldsymbol{\omega}+2)\end{aligned}\tag{5.59}$$

where λ and μ are real-valued Lagrange multipliers with μ being arbitrary and λ satisfying $\boldsymbol{C}_r^H\boldsymbol{R}_r\boldsymbol{C}_r+\lambda\boldsymbol{R}_l>0$ so that $f_1(\boldsymbol{\omega},\lambda,\mu)$ can be minimized with respect to $\boldsymbol{\omega}$. Then,

$$f_1(\boldsymbol{\omega},\lambda,\mu)\leqslant\boldsymbol{C}_r^H\boldsymbol{R}_r\boldsymbol{C}_r,\ \text{for}\quad\boldsymbol{\omega}\in\mathscr{S}\tag{5.60}$$

with equality on the boundary of $\mathscr{S}$.

Consider the following condition:

$$\frac{\boldsymbol{a}^H\left(\boldsymbol{C}_r^H\boldsymbol{R}_r\boldsymbol{C}_r\right)^{-2}\boldsymbol{a}}{\left(\boldsymbol{a}^H\left(\boldsymbol{C}_r^H\boldsymbol{R}_r\boldsymbol{C}_r\right)^{-1}\boldsymbol{a}\right)^2}\le\epsilon^2\tag{5.61}$$

When this condition is satisfied, the solution to $\mathscr{P}$ can be expressed by

$$\boldsymbol{\omega}=\frac{\left(\boldsymbol{C}_r^H\boldsymbol{R}_r\boldsymbol{C}_r\right)^{-1}\boldsymbol{a}}{\boldsymbol{a}^H\left(\boldsymbol{C}_r^H\boldsymbol{R}_r\boldsymbol{C}_r\right)^{-1}\boldsymbol{a}}\tag{5.62}$$

which satisfies the ambiguity constraint in $\mathscr{P}$.

Otherwise, we note that (5.60) can be written as

$$\begin{aligned}f_1(\boldsymbol{\omega},\lambda,\mu)&=\left(\boldsymbol{\omega}-\mu(\boldsymbol{C}_r^H\boldsymbol{R}_r\boldsymbol{C}_r+\lambda\boldsymbol{R}_0)^{-1}\boldsymbol{a}\right)^H\\ &\cdot(\boldsymbol{C}_r^H\boldsymbol{R}_r\boldsymbol{C}_r+\lambda\boldsymbol{R}_0)\left(\boldsymbol{\omega}-\mu(\boldsymbol{C}_r^H\boldsymbol{R}_r\boldsymbol{C}_r+\lambda\boldsymbol{R}_0)^{-1}\boldsymbol{a}\right)\\ &-\mu^2\boldsymbol{a}^H(\boldsymbol{C}_r^H\boldsymbol{R}_r\boldsymbol{C}_r+\lambda\boldsymbol{R}_0)^{-1}\boldsymbol{a}-\lambda\epsilon^2+2\mu\end{aligned}\tag{5.63}$$

Hence, the unconstrained minimizer of $f_1(\boldsymbol{\omega}, \lambda, \mu)$, for a fixed λ and μ, is given by

$$\tilde{\boldsymbol{\omega}} = \mu(\boldsymbol{C}_r^H \boldsymbol{R}_r \boldsymbol{C}_r + \lambda \boldsymbol{R}_0)^{-1}\boldsymbol{a} \tag{5.64}$$

Clearly, we have

$$\begin{aligned} f_2(\lambda, \mu) = f_1(\tilde{\boldsymbol{\omega}}, \lambda, \mu) = &- \mu^2 \boldsymbol{a}^H (\boldsymbol{C}_r^H \boldsymbol{R}_r \boldsymbol{C}_r + \lambda \boldsymbol{R}_0)^{-1}\boldsymbol{a} \\ &- \lambda\epsilon^2 + 2\mu \end{aligned} \tag{5.65}$$

The maximization of $f_2(\lambda, \mu)$ with respect to μ gives

$$\tilde{\mu} = \frac{1}{\boldsymbol{a}^H (\boldsymbol{C}_r^H \boldsymbol{R}_r \boldsymbol{C}_r + \lambda \boldsymbol{R}_0)^{-1}\boldsymbol{a}} \tag{5.66}$$

and

$$f_3(\lambda) = f_2(\lambda, \tilde{\mu}) = -\lambda\epsilon^2 + \frac{1}{\boldsymbol{a}^H (\boldsymbol{C}_r^H \boldsymbol{R}_r \boldsymbol{C}_r + \lambda \boldsymbol{R}_0)^{-1}\boldsymbol{a}} \tag{5.67}$$

The maximization of the above function with respect to λ gives

$$\frac{\boldsymbol{a}^H (\boldsymbol{C}_r^H \boldsymbol{R}_r \boldsymbol{C}_r + \tilde{\lambda} \boldsymbol{R}_0)^{-2}\boldsymbol{a}}{\left(\boldsymbol{a}^H (\boldsymbol{C}_r^H \boldsymbol{R}_r \boldsymbol{C}_r + \tilde{\lambda} \boldsymbol{R}_0)^{-1}\boldsymbol{a}\right)^2} = \epsilon^2 \tag{5.68}$$

It can be observed that when the condition in (5.61)is not satisfied, (5.68) admits a unique solution $\lambda > 0$. Moreover, the left-hand side of (5.68) is a monotonically decreasing function of λ, allowing λ to be efficiently determined using methods such as Newton's method. Equation (5.68) can also be reformulated as a degree- M polynomial with a single unknown, from which a unique positive solution λ can be obtained.

Note that using (5.66) in (5.64) yields

$$\tilde{\boldsymbol{\omega}} = \frac{(\boldsymbol{C}_r^H \boldsymbol{R}_r \boldsymbol{C}_r + \tilde{\lambda} \boldsymbol{R}_0)^{-1}\boldsymbol{a}}{\boldsymbol{a}^H (\boldsymbol{C}_r^H \boldsymbol{R}_r \boldsymbol{C}_r + \tilde{\lambda} \boldsymbol{R}_0)^{-1}\boldsymbol{a}} \tag{5.69}$$

This approach yields a solution to the azimuth-ambiguity-constrained optimization problem $\mathscr{P}$. Under a fixed upper bound on azimuth-ambiguity power ϵ^2 with respect to $\tilde{\boldsymbol{\omega}}$, the minimum range-ambiguous signal power can thus be achieved.

Conversely, the range-ambiguous signal power may be imposed as a constraint, allowing us to minimize the azimuth-ambiguity power of the desired signal; the corresponding optimization model can thus be expressed as follows.

$$\mathscr{P}_c : \begin{cases} \min\limits_{\omega(f_\eta)} & \omega^H(f_\eta)\boldsymbol{R}_0(f_\eta)\omega(f_\eta) \\ \text{s.t.} & \omega^H(f_\eta)\boldsymbol{a}(f_\eta) = 1 \\ & \omega^H(f_\eta)\boldsymbol{C}_r^H\boldsymbol{R}_r(f_\eta)\boldsymbol{C}_r\omega(f_\eta) \le \epsilon_c^2(f_\eta) \end{cases} \tag{5.70}$$

The solution to $\mathscr{P}_c$ produces results similar to those of problem $\mathscr{P}$, and in fact, the solutions of $\mathscr{P}$ and $\mathscr{P}_c$ are mutually convertible.

Since the azimuth-ambiguity-constrained optimization problem $\mathscr{P}$ explicitly accounts for the range-ambiguous signal power, the resulting solution provides improved imaging performance compared with the conventional DBF reconstruction filter.

5.3.2.3 Specific PRFs

When the PRF f_p belongs to one of the specific PRFs, the time between successive pulses can be given by

$$T = \frac{1}{f_p} = \frac{1}{\text{PRF}^{\text{SPEC}}} = \frac{Kd_a}{2V_r} \tag{5.71}$$

where $K \in \{1, \ldots, M-1\}$. In this case, whatever the filter $\omega(f)$ is, there always exists

$$\omega^H(f_\eta)\boldsymbol{a}_k(f_\eta) = \omega^H(f_\eta)\boldsymbol{a}_{k+K}(f_\eta) \tag{5.72}$$

Thus, the reconstructible Doppler bandwidth cannot be beyond $2V_r/d_a$. The space-time sampling pattern with respect to a specific time interval T is depicted in Fig. 5.9.

From Fig. 5.9, the impulse response of the $(j+K)$th channel is accordingly given by

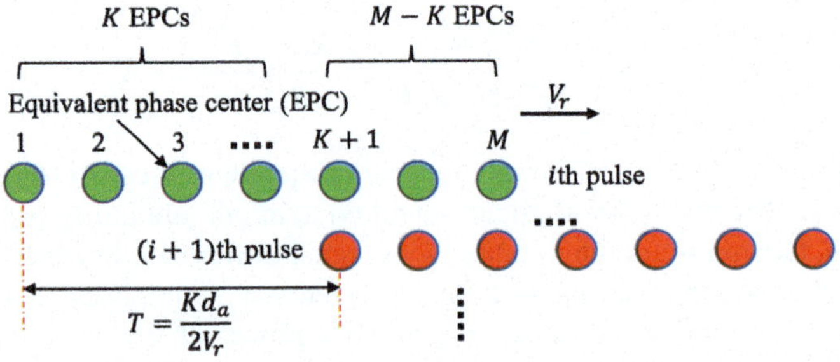

Fig. 5.9 Space-time sampling patterns with respect to one specific PRF

$$s_{j+K}^0(\eta) = s^0(\eta - \frac{(K+j-1)d_a}{2V_r}) = s^0(\eta - \frac{j-1}{\mathrm{PRI}} - \mathrm{PRI}) \tag{5.73}$$

The corresponding sampled signal of $s_{j+K}^0(\eta)$ with $\eta = n\mathrm{PRI}, n = 1, 2, \ldots$ can be written as

$$s_{j+K}^0[n] = s^0[n\mathrm{PRI} - \frac{j-1}{\mathrm{PRI}} - \mathrm{PRI}] \tag{5.74}$$

Moreover, the impulse response of the jth channel is given by

$$s_j^0(\eta) = s^0\left(\eta - \frac{(j-1)d_a}{2V_r}\right) = s^0\left(\eta - \frac{j-1}{\mathrm{PRI}}\right) \tag{5.75}$$

The corresponding sampled impulse response of $s_j^0(t)$ is written as

$$s_j^0(n) = s^0\left(n\mathrm{PRI} - \frac{j-1}{\mathrm{PRI}}\right) \tag{5.76}$$

The $(j+K)$th channel exhibits the same impulse response as the jth channel when the latter is shifted by one pulse-repetition interval T (T-shift operation). In other words, the final $M-K$ equivalent phase centers (EPCs) of the received signal coincide with the first $M-K$ EPCs recorded in the subsequent pulse-repetition interval.

When the APC technique is applied in a multichannel SAR configuration using particular PRFs, the desired signal in the last $M-K$ channels matches that in the first $M-K$ channels, whereas the range-ambiguous components in the last $M-K$ channels exhibit opposite polarity relative to those in the first $M-K$ channels.

This property enables the combination of the echoes from the last $M-K$ channels with the T-shift echoes from the first $M-K$ channels so that range ambiguity can be mitigated and SNR enhanced simultaneously. The detailed procedure used for range-ambiguity suppression is outlined below.

We will consider two cases: (1) $M-K \geq K$ and (2)$M-K < K$.

$M-K \geq K$

In this case, the first-order range ambiguity can be totally removed by a superposition operation. Moreover, extra DOFs can be used to minimize the noise level. In this case, with the help of equation (5.39), the azimuth spectrum of the signal of interest and the range-ambiguous signal in vector format will be reformulated as

$$\tilde{\boldsymbol{S}}^0(f_\eta) = \boldsymbol{A}_p(f_\eta)\boldsymbol{S}_p^0(f_\eta) + \boldsymbol{n}(f_\eta) \tag{5.77}$$

and

$$\tilde{\boldsymbol{S}}^r(f_\eta) = \boldsymbol{C}_r\boldsymbol{A}_p(f_\eta)\boldsymbol{S}_p^r(f_\eta) + \boldsymbol{n}(f_\eta) \tag{5.78}$$

where $\boldsymbol{A}_p(f_\eta) = [a_1, a_2, \ldots, a_K]$, $\boldsymbol{S}_p^0(f_\eta)$ and $\boldsymbol{S}_p^r(f_\eta)$ depend on $\boldsymbol{S}^0(f)$ and $\boldsymbol{S}^r(f_\eta)$, respectively.

Due to the condition $M - K \geq K$, one can obtain the following optimization problem $\mathscr{P}_1$.

$$\mathscr{P}_1 : \begin{cases} \min\limits_{\boldsymbol{\omega}(f_\eta)} & \boldsymbol{\omega}^H(f_\eta)\boldsymbol{\omega}(f_\eta) \\ \text{s.t.} & \boldsymbol{\omega}^H(f_\eta)\boldsymbol{A}_p(f_\eta) = \boldsymbol{u}_k^T \\ & \boldsymbol{\omega}^H(f_\eta)\boldsymbol{C}_r\boldsymbol{A}_p(f_\eta) = \boldsymbol{0} \end{cases} \tag{5.79}$$

where $\boldsymbol{u}_k = [0, 0, \ldots, 1, \ldots, 0]$ is a $K \times 1$ vector of zeros except for the kth entry, which is equal to one. The first constraint in $\mathscr{P}_1$ ensures the perfect reconstruction of the signal's Doppler spectrum of interest, and the second constraint in $\mathscr{P}_1$ shows that the first-order range ambiguity can be removed.

Let

$$\boldsymbol{B}(f_\eta) = [\boldsymbol{A}_p(f_\eta), \boldsymbol{C}_r\boldsymbol{A}_p(f_\eta)] \tag{5.80}$$

where $\boldsymbol{B}(f_\eta)$ is an $M \times 2K$ matrix. The rank of matrix $\boldsymbol{B}(f_\eta)$ can be calculated as r (obviously, $r \geq K$), then remove the last $2K - r$ columns of $\boldsymbol{B}(f_\eta)$ to obtain the matrix $\boldsymbol{B}_r(f_\eta) \in \boldsymbol{B}(f_\eta)$. After that, based on the method of Lagrange multipliers, the solution of the optimization $\mathscr{P}_1$ is given by

$$\boldsymbol{\omega}(f_\eta) = \boldsymbol{B}_r(f_\eta)\big(\boldsymbol{B}_r^H(f_\eta)\boldsymbol{B}_r(f_\eta)\big)^{-1}\boldsymbol{v}_k \tag{5.81}$$

where $\boldsymbol{v}_k = [\boldsymbol{u}_k, \boldsymbol{0}]$ is a $r \times 1$ vector.

$M - K < K$

In this case, $2K > M$ so that $\boldsymbol{B}_r^H(f_\eta)\boldsymbol{B}_r(f_\eta)$ is irreversible, and the number of DOFs K is not enough to totally remove the first-order range ambiguity. Therefore, an optimization model to minimize the range ambiguity is established as follows:

$$\mathscr{P}_2 : \begin{cases} \min\limits_{\boldsymbol{\omega}(f_\eta)} & \boldsymbol{\omega}^H(f_\eta)\boldsymbol{C}_r\boldsymbol{R}_r(f_\eta)\boldsymbol{C}_r^H\boldsymbol{\omega}(f_\eta) \\ \text{s.t.} & \boldsymbol{\omega}^H(f_\eta)\boldsymbol{A}_p(f_\eta) = \boldsymbol{u}_k^T \end{cases} \tag{5.82}$$

Based on the method of Lagrange multipliers, the solution of the optimization problem $\mathscr{P}_2$ is accordingly given by

$$\boldsymbol{\omega}(f_\eta) = \boldsymbol{R}_r^{-1}(f_\eta)\boldsymbol{A}_p(f_\eta)\big(\boldsymbol{A}_p^H(f_\eta)\boldsymbol{R}_r^{-1}(f_\eta)\boldsymbol{A}_p(f_\eta)\big)^{-1}\boldsymbol{u}_k \tag{5.83}$$

where the matrix $\boldsymbol{R}_r$ contains a diagonal loading factor so that $\boldsymbol{R}_r$ is invertible.

5.4 Reconstruction Performance of the Proposed Algorithm

The imaging performance of an APC-enabled multichannel SAR system with a uniform PRF has already been analyzed in [21]. Hence, this section focuses on the case of a nonuniform PRF and evaluates the reconstruction performance of the proposed approach using the representative spaceborne HRWS SAR parameters listed in Table 5.4.

5.4.1 Reconstruction Performance Versus PRF

We begin by showing the reconstructed spectra of both the desired signal and the range-ambiguous signal under a nonuniform spatial sampling PRF 1350 Hz, with the simulation settings summarized in Table 5.4.

For a comparison, the azimuth patterns between the 4-channel and 8-channel SAR systems are identical with the same Doppler bandwidth $B_d = 3700$ Hz, as shown in Fig. 5.10.

Figure 5.11 presents the reconstructed spectra of the desired and range-ambiguous signals using the DBF reconstruction method. Without the APC technique, the reconstructed spectra of the range-ambiguous signal and the desired signal become identical. Once APC is incorporated into the multichannel SAR system, however, one can observe that the range-ambiguous energy is attenuated by the DBF reconstruction. The 4-channel and 8-channel reconstructed results are shown in Fig. 5.11a and b, respectively. As shown in Fig. 5.11, the effective PRFs of the 4-channel and 8-channel systems 5400 Hz and 800 Hz, respectively. Additional distinctions between these two systems are summarized in Table 5.5. Consequently, the azimuth-ambiguity level of the 8-channel system is lower than that of the 4-channel configuration, and its range-ambiguity level is also reduced. Thus, with a fixed Doppler bandwidth, the

Table 5.4 System parameters

Parameter	Value
Doppler bandwidth	3700 Hz
Radar velocity	7596 m/s
Carrier frequency	5.4 GHz
Total receiver length	12 m
Length of transmitting antenna	4.5 m
Number of receivers	4/8
Subaperture length	3 m/1.5 m
Non-uniform PRF	1350 Hz
PRF range	1300–1500 Hz

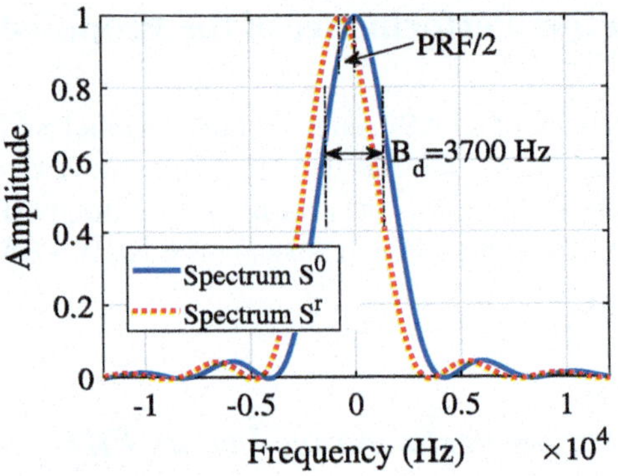

Fig. 5.10 Azimuth patterns in the signal of interest $S^0(f_\eta)$ and range-ambiguous signal $S^r(f_\eta)$

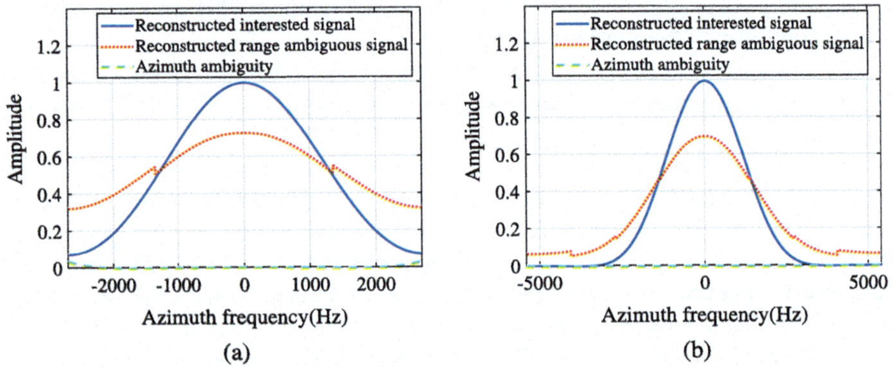

Fig. 5.11 Reconstructed spectrum of the signal of interest and range-ambiguous signal by the DBF reconstruction method. **a** $M = 4$. **b** $M = 8$

Table 5.5 Performance comparison with different methods

Method	AASR	APC gain
4-channel (MI)	−24.90 dB	1.13
8-channel (MI)	−34.42 dB	1.22
4-channel (Proposed)	−22.85 dB	1.21
8-channel (Proposed)	−21.30 dB	1.81

DBF-based APC/multichannel SAR arrangement yields decreasing range ambiguity as the number of channels M increases.

Nevertheless, the extent of range-ambiguity suppression achieved with the DBF reconstruction method is limited. When the proposed reconstruction approach is applied to the APC/multichannel system, the resulting spectra are shown in Fig. 5.12 and a detailed comparison of the 4-channel and 8-channel performance is again provided in Table 5.5. For $M = 4$, the results in Fig. 5.12a indicate only a marginal

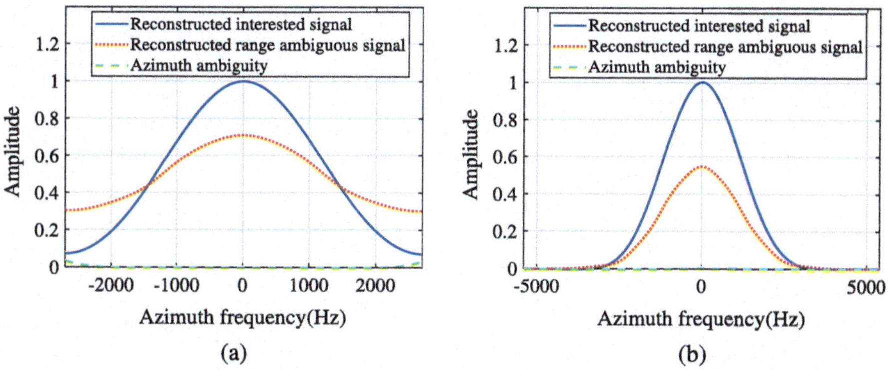

Fig. 5.12 Reconstructed spectrum of the signal of interest and range-ambiguous signal by the proposed method. **a** $M = 4$. **b** $M = 8$

improvement relative to Fig. 5.11a. For $M = 8$, the results in Fig. 5.12b show a pronounced enhancement of range-ambiguity suppression compared with Fig. 5.11b. Therefore, the proposed reconstruction method delivers substantially better range-ambiguity mitigation than DBF reconstruction when used within an APC-enabled multichannel SAR system.

To further examine the reconstruction capability of the proposed method, the PRF is varied within the interval 1300–2000 Hz. Again, both 4-channel and 8-channel systems are considered. Under these conditions, the corresponding effective PRF ranges are $4200 \sim 8000$ Hz and $10400 \sim 16000$ Hz, respectively. For the method proposed, $\tilde{\lambda} = 14.28$ and $\tilde{\lambda} = 7.69$ are assigned to construct the reconstruction filters $\tilde{\omega}$ for the 4-channel and 8-channel systems, according to equation (5.69). These values of $\tilde{\lambda}$ are chosen to maintain the low azimuth ambiguity level below -20 dB. As $\tilde{\lambda} \to \infty$, the filter approaches the minimum-azimuth-ambiguity design, whereas $\tilde{\lambda} \to 0$ yields minimum range ambiguity but unacceptably high azimuth ambiguity in the desired signal. After applying the reconstruction filters, the resulting spectra are finally passed through a bandpass filter $\mathrm{rect}(f/B_d)$ to enforce the desired Doppler bandwidth.

Figure 5.13 plots the APC gain G_{APC}, the AASR, and the SNR as functions of PRF for two reconstruction approaches: the proposed method (green curves) and the DBF-based reconstruction (red curves). The left and right columns correspond to the 4-channel and 8-channel SAR configurations, respectively. As illustrated in Fig. 5.13a and b, the proposed method consistently delivers higher APC gains. When the DBF reconstruction is used in a 4-channel system, the APC gain never exceeds about 0.55 dB and even drops below -10 dB near certain PRFs, resulting in severe range ambiguity; thus, the DBF approach becomes unsuitable for APC-based multichannel SAR in such cases. Although the DBF method achieves slightly higher APC gains in the 8-channel case than in the 4-channel case, the maximum gain remains below 1.2 dB, offering only about 1–2 dB of range-ambiguity reduction. In contrast, the

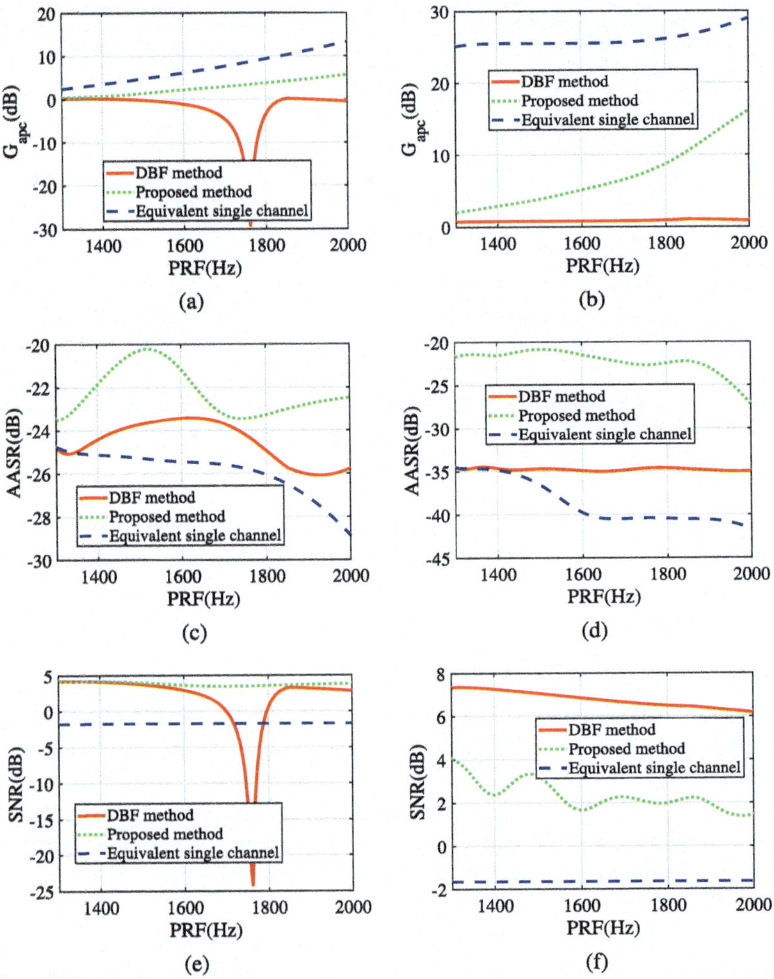

Fig. 5.13 APC gain G_{APC}, the AASR and SNR versus the PRF for the 4-channel (left) and 8-channel (right) SAR systems with parameters given in Table 5.4 when the APC technique is used. The green and red lines represent the proposed reconstruction method and DBF reconstruction method, respectively

proposed method attains peak gains of about 6 dB and 15 dB for the 4- and 8-channel systems, respectively.

Figure 5.13c and d compare the AASR obtained by the two reconstruction strategies. One can observe that the DBF method has a better AASR than the proposed method because the proposed method achieves a high APC gain at the expanse of the azimuth ambiguity. Nevertheless, the AASR of the proposed method is still less than −20 dB, which is sufficient to satisfy the requirement of the SAR image. Furthermore, the SNR of different methods is discussed in Fig. 5.13e and f. The DBF

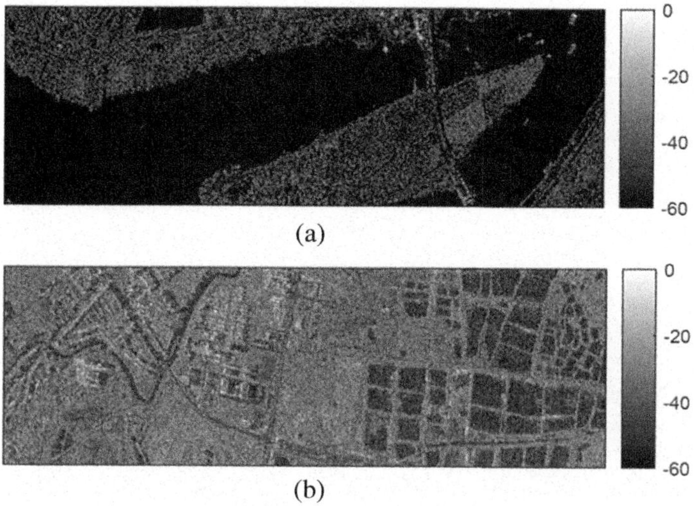

Fig. 5.14 The scene of interest and range ambiguity are provided as input for the echo simulation. The horizontal direction is the azimuth direction, and the vertical direction is the range direction. **a** The scene of interest. **b** Range-ambiguity scene

method would have a poor SNR for highly nonuniform spatial sampling, while the proposed method can avoid this issue. In addition, the proposed method achieves a lower SNR than the DBF method but only by approximately 4 dB. Therefore, the SNR of the proposed method is acceptable for SAR images.

5.4.2 Distributed Scene Simulation

To further demonstrate the benefits and effectiveness of the proposed reconstruction method, spaceborne APC/multichannel SAR simulation experiments are conducted. Under a strip-map imaging geometry, the simulation parameters (listed in Table 5.4 use $PRF = 2000$ Hz and $M = 8$). Figure 5.14 shows both the desired scene and the corresponding range-ambiguity scene (with approximately −15 dB relative power), which serve as input to the echo simulator. Figure 5.15a shows the reconstructed image for a multichannel SAR system without APC, where the image quality is clearly poor. Applying the DBF reconstruction to the APC/multichannel SAR system yields the results in Fig. 5.15b, showing only modest improvement. One can see that the image quality has slightly improved, but it is still very poor. In contrast, the proposed method (Fig. 5.15c) effectively suppresses the range-ambiguity energy. The quantified range-ambiguity levels for the region marked by red circles in Fig. 5.15 are summarized in Table 5.6.

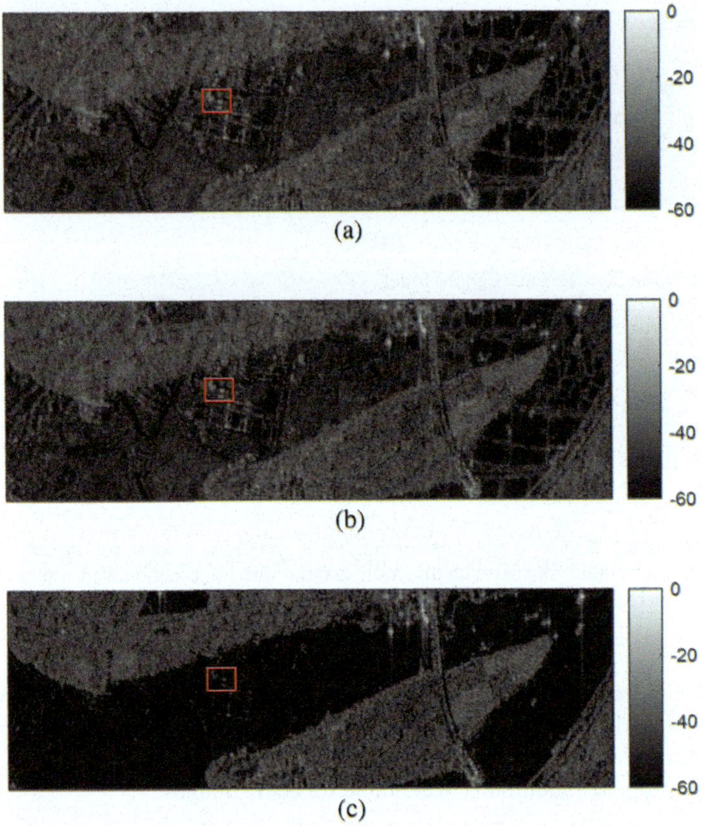

Fig. 5.15 Distributed scene simulation results for the APC/multichannel SAR system. **a** The multichannel SAR system without the APC technique. **b** The APC/multichannel SAR system with the DBF reconstruction method. **c** The APC/multichannel SAR system with the proposed method

Table 5.6 Range ambiguity level after different methods

Method	No APC	APC (DBF)	APC (Proposed)
Range ambiguity level	−20.71 dB	−22.67 dB	−35.31 dB

5.5 Summary

The APC technique has been extensively examined for range-ambiguity mitigation in both conventional and multichannel SAR systems due to its low implementation cost and its effectiveness for suppressing both point-target and distributed ambiguities. However, most previous studies have considered APC within the framework of DBF-on-receive multichannel SAR, where the DBF reconstruction does not explicitly account for the suppression of range ambiguity–even though, under APC, the azimuth spectrum of the ambiguous signal differs from that of the desired signal. This work introduces an enhanced reconstruction method for APC/multichannel SAR based on a quadratically constrained optimization formulation, which incorporates the modified azimuth-spectral characteristics of the range-ambiguous signal induced by APC. Unlike the DBF approach, the proposed method enables minimizing range ambiguity while maintaining a specified level of azimuth ambiguity. Moreover, for certain PRF configurations, the method can cancel the odd-order range-ambiguous components entirely.

References

1. Cumming IG, Wong FH. Digital processing of synthetic aperture radar data. Artech house. 2005;1(3).
2. Yun-kai D, Feng-jun Z. Wang Yu. Brief analysis on the development and application of spaceborne SAR. Journal of Radars. 2012;1(1):1–10.
3. Curlander JC, McDonough RN. Synthetic aperture radar. vol. 11. Wiley, New York; 1991.
4. Elachi C. Spaceborne radar remote sensing: applications and techniques. New York, IEEE Press, 1988, 285 p. 1988;.
5. Freeman A, Johnson WT, Huneycutt Bea, et al. The" myth" of the minimum SAR antenna area constraint. IEEE Transactions on Geoscience and Remote Sensing. 2000;38(1):320–324.
6. Claassen J, Eckerman J. A system concept for wide swath constant incident angle coverage. 1978;.
7. Fischer C, Heer C, Krieger G, et al. A high resolution wide swath SAR. In: Proc. EUSAR. vol. 6; 2006. .
8. Gebert N, Krieger G, Moreira A. High resolution wide swath SAR imaging with digital beamforming-performance analysis, optimization, system design. In: European Conference on Synthetic Aperture Radar (EUSAR). VDE Verlag GmbH; 2006. p. 4.
9. Heer C, Soualle F, Zahn R, et al. Investigations on a new high resolution wide swath SAR concept. In: IGARSS 2003. 2003 IEEE International Geoscience and Remote Sensing Symposium. Proceedings (IEEE Cat. No. 03CH37477). vol. 1. IEEE; 2003. p. 521–523.
10. Bordoni F, Younis M, Gebert N, et al. Performance investigation on the high-resolution wide-swath SAR system with monostatic architecture. In: 8th European Conference on Synthetic Aperture Radar. VDE; 2010. p. 1–4.
11. Li F, Johnson W. Ambiguities in spacebornene synthetic aperture radar systems. IEEE Transactions on Aerospace and Electronic Systems. 1983;(3):389–397.
12. Griffiths H, Mancini P. Ambiguity suppression in SARs using adaptive array techniques. In: IGARSS'91; Proceedings of the 11th Annual International Geoscience and Remote Sensing Symposium. vol. 2; 1991. p. 1015–1018.
13. Gebert N, Krieger G, Moreira A. Digital beamforming on receive: Techniques and optimization strategies for high-resolution wide-swath SAR imaging. IEEE Transactions on Aerospace and Electronic Systems. 2009;45(2).
14. Gebert N. Multi-channel azimuth processing for high-resolution wide-swath SAR imaging [Ph. D. Dissertation]. Universitat Karlsruhe, Karlsruhe, Germany. 2009;.
15. Krieger G, Gebert N, Moreira A. Unambiguous SAR signal reconstruction from nonuniform displaced phase center sampling. IEEE Geoscience and Remote Sensing Letters. 2004;1(4):260–264.
16. Huber S, Younis M, Patyuchenko A, et al. Digital beam forming techniques for spaceborne reflector SAR systems. In: 8th European Conference on Synthetic Aperture Radar. VDE; 2010. p. 1–4.

17. Cerutti-Maori D, Sikaneta I, Klare J, et al. MIMO SAR processing for multichannel high-resolution wide-swath radars. IEEE Transactions on Geoscience and Remote Sensing. 2013;52(8):5034–5055.
18. Sikaneta I, Gierull CH, Cerutti-Maori D. Optimum signal processing for multichannel SAR: With application to high-resolution wide-swath imaging. IEEE Transactions on Geoscience and Remote Sensing. 2014;52(10):6095–6109.
19. Wang W, Wang R, Deng Y, et al. Azimuth ambiguity suppression with an improved reconstruction method based on antenna pattern for multichannel synthetic aperture radar systems. IET Radar, Sonar & Navigation. 2015;9(5):492–500.
20. Dall J, Kusk A. Azimuth phase coding for range ambiguity suppression in SAR. In: IGARSS 2004. 2004 IEEE International Geoscience and Remote Sensing Symposium. vol. 3. IEEE; 2004. p. 1734–1737.
21. Bordoni F, Younis M, Krieger G. Ambiguity suppression by azimuth phase coding in multichannel SAR systems. IEEE Transactions on Geoscience and Remote Sensing. 2011;50(2):617–629.
22. Papoulis A. Signal analysis. vol. 191. McGraw-Hill New York; 1977.
23. Hudson J. Adaptive array principles. vol. 11. IET; 1981.
24. Tian Z, Bell KL, Van Trees HL. A recursive least squares implementation for LCMP beamforming under quadratic constraint. IEEE Transactions on Signal Processing. 2001;49(6):1138–1145.

Chapter 6
Ambiguity Suppression Scheme for Quad-Pol SAR Based on Quasi-Orthogonal Waveforms

6.1 Introduction

The design of radar waveforms has been an active research topic for many years [1–3], as it fundamentally governs the achievable performance of radar systems. Quasi-orthogonal waveforms have emerged as an effective approach for mitigating undesired ambiguity energy in Synthetic Aperture Radar (SAR) systems [4, 5]. Nevertheless, achieving waveform orthogonality in SAR is substantially more challenging than in communication systems, mainly because of the following factors:

- *Distributed target scene.*
- *Imaging performance.*

The Cross-Correlation Energy (CCE) observed between waveforms after range compression tends to appear smeared or shifted to different range–azimuth locations [6, 7]. Consequently, employing quasi-orthogonal waveforms with inherently low CCE provides a practical means to alleviate ambiguity effects. Nevertheless, based on Parseval's theorem, Krieger [8] noted that waveforms characterized by flat spectra—such as up-down chirps—yield a constant CCE value for any waveform pair, which suggests that non-flat spectral designs are required to achieve further CCE reduction. Moreover, because SAR functions as an imaging sensor, the transmitted waveforms must maintain low Peak-to-Side-Lobe Ratios (PSLRs) and Integrated Side-Lobe Ratios (ISLRs) to guarantee adequate imaging quality.

6.1.1 Background and Previous Works

Quadrature-polarimetric (quad-pol) SAR, which targets the retrieval of environmental scattering characteristics from all four polarimetric channels (HH, HV, VH, VV), has become a crucial instrument in remote sensing [9–11]. Current quad-pol SAR implementations primarily rely on three transmission strategies:

R. Wang and W. Wang, *Synthetic Aperture Radar (SAR) Waveform Design and Signal Processing*, https://doi.org/10.1007/978-981-95-7549-7_6

(1) Conventional quad-pol SAR, where H- and V-polarized pulses are transmitted alternately [12];

(2) Hybrid quad-pol SAR, which sequentially transmits left- and right-circular polarizations [13];

(3) $\pm\pi/4$ quad-pol SAR, transmitting (H–V) and (H+V) polarizations in an alternating manner [14].

At the receiver, echoes generated under these transmission schemes are recorded on both H and V channels. For a conventional quad-pol system, the complete scattering matrix can be directly reconstructed, yet this mode is known to suffer from significant range ambiguities in the cross-polarized channels (HV and VH) [12]. In contrast, hybrid and $\pm\pi/4$ quad-pol SARs recover the full scattering matrix (HH, HV, VH, VV) via a straightforward mathematical transformation applied to the measured data. These two modes also achieve a more favorable ratio between desired signals and range-ambiguous components, thereby substantially reducing range ambiguities in HV and VH [12]. However, previous studies have shown that both hybrid and $\pm\pi/4$ configurations may experience pronounced azimuth ambiguities in the cross-polarized channels [15].

Azimuth Phase Coding (APC) [16] operates by applying an azimuth-dependent phase modulation to the transmitted pulses and performing the corresponding demodulation on reception. This modulation introduces alternating 0 and π phases for the odd-indexed range ambiguous components [16], effectively shifting them within the Doppler frequency domain. As a result, APC converts range ambiguities into azimuth ambiguities through transmit-side phase modulation, allowing the resulting ambiguity energy to be suppressed in the azimuth dimension. Using this strategy, prior studies have demonstrated that APC can effectively mitigate range ambiguities in both conventional SAR and conventional quad-pol SAR architectures [17]. In parallel, another line of research exploits quasi-orthogonal waveform transmission [4, 5, 18] to suppress range ambiguities without requiring additional system resources such as extra receive channels [19]. Unlike APC, whose suppression mechanism relies on Doppler shifting, the alternating-transmission scheme based on quasi-orthogonal signals achieves ambiguity reduction by capitalizing on the low CCE properties of the employed waveforms. Although quasi-orthogonal waveforms have been extensively examined for mitigating range ambiguities in conventional SAR systems [4, 5], their potential extension to quad-pol SAR has not yet been thoroughly investigated.

Given the various design constraints associated with employing quasi-orthogonal waveforms in SAR systems, Nonlinear Frequency Modulation (NLFM) signals have emerged as strong candidates. Earlier studies on NLFM signals primarily addressed their synthesis [20, 21] and optimization strategies [22, 23]. The practical realization of NLFM waveforms exhibiting low sidelobe levels has been examined in [24–26]. However, these approaches mainly concentrate on improving the auto-correlation performance and do not explicitly consider correlation properties among different signals. Furthermore, a high-accuracy NLFM signal generator—capable of predistortion compensation—was developed and deployed in the LuTan-1 (LT-1) mission [27].[1] The demonstrated effectiveness of NLFM signal transmission in SAR therefore motivates this work to focus on the development of quasi-orthogonal NLFM waveforms.

6.1.2 Main Contribution and Organization

This work focuses on the development and practical realization of quasi-orthogonal NLFM waveforms aimed at suppressing ambiguities in quad-pol SAR systems. The primary contributions are as follows:

- Design of quasi-orthogonal NLFM waveforms that simultaneously achieve low CCE and low PSLRs. The resulting metrics satisfy the imaging requirements even without windowing, thereby preventing any associated SNR degradation.
- The Range and Azimuth Ambiguity-to-Signal Ratios (RASR and AASR) of quad-pol SAR systems were analyzed in [15]. However, because that study did not incorporate the impact of varying transmitted waveforms, this work aims to bridge that gap by presenting more general expressions for evaluating AASR and RASR that explicitly account for waveform diversity.
- Building upon the developed quasi-orthogonal NLFM waveforms, this study introduces a novel intertwined polarimetric transmission strategy, together with its APC-enabled extension, to further enhance ambiguity suppression capability.

The structure of this paper is outlined as follows. Section 6.2 describes the signal formulation together with the optimization strategy adopted for designing quasi-orthogonal NLFM waveforms. Section 6.3 details how these waveforms are integrated with the APC technique in a conventional quad-pol SAR configuration. Comprehensive simulation studies supporting the theoretical developments are presented in Sect. 6.4. Finally, Sect. 6.5 provides concluding remarks.

Notation: Bold uppercase and lowercase letters represent matrices and column vectors, respectively. The symbol '$\leq$' is interpreted elementwise, $(\cdot)^*$ denotes the complex conjugate, and $\odot$ stands for the Hadamard product.

[1] The LuTan-1 (LT-1) mission, approved in 2016 with an original launch target of 2020, forms a key element of China's long-term civil space program. LT-1 is equipped for spaceborne L-band interferometric observations, offering capabilities such as differential interferometry and full quad-polarization. It supports multiple polarization architectures, including conventional, hybrid, $\pm\pi/4$ quad-pol, and compact-pol modes. The mission is intended to generate high-accuracy digital elevation models and to provide remote-sensing products for applications such as biomass estimation, disaster monitoring, and climate and environmental studies.

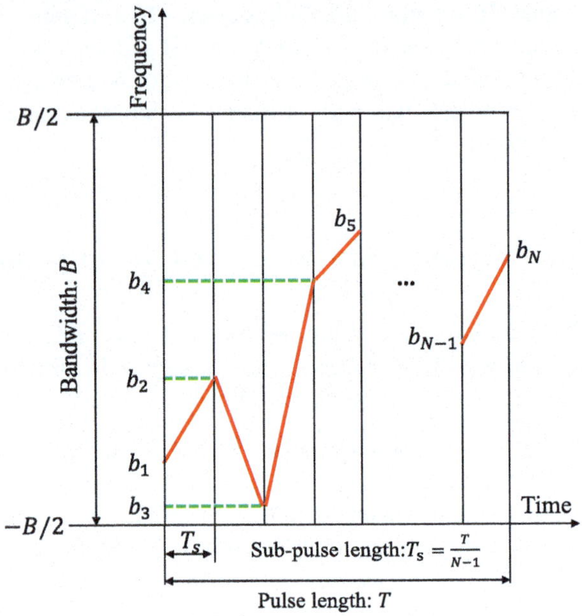

Fig. 6.1 Illustration of the instantaneous frequency history

6.2 The Design of Quasi-Orthogonal NLFM Waveforms

6.2.1 Signal Model

This section develops the NLFM waveform by concatenating multiple LFM chirp sub-pulses, as illustrated in Fig. 6.1. In contrast to Orthogonal Frequency-Division Multiplexing (OFDM) chirp designs [28], the instantaneous frequency of the proposed NLFM signal evolves smoothly along a continuous, piecewise linear profile.

Let B and T represent the bandwidth and pulse duration of the NLFM waveform, respectively. The signal can then be synthesized using $N-1$ chirp sub-pulses, as

$$\bar{s}(t) = \sum_{n=1}^{N-1} s_n(t-(n-1)T_s), \tag{6.1}$$

where

$$s_n(t) = \text{rect}\left(\frac{t-T_s/2}{T_s}\right)\exp(j\pi\alpha_n t^2)\exp(j2\pi b_n t), \tag{6.2}$$

with

$$\text{rect}\,(t) = \begin{cases} 1, & -0.5 \le t \le 0.5 \\ 0, & \text{elsewhere} \end{cases}, \tag{6.3}$$

where $T_s = \frac{T}{N-1}$ denotes the duration of each sub-pulse, and α_n and b_n correspond to the FM slope and frequency offset of the nth sub-pulse, respectively. Specifically, to ensure continuity of the instantaneous frequency

$$\alpha_n = \frac{b_{n+1} - b_n}{T_s}, \tag{6.4}$$

where b_n and b_{n+1} represent the initial and final instantaneous frequencies of the nth sub-pulse, for $n = 1, \ldots, N-1$. Clearly, the resulting NLFM waveform is determined by the control point vector $\boldsymbol{b} = (b_1, b_2, \ldots, b_N)$, which will be collectively denoted as $\bar{s}(t; \boldsymbol{b})$.

6.2.2 *Correlation Function*

The cross-correlation function between two NLFM waveforms, $\bar{s}_1(t; \boldsymbol{b}_1)$ and $\bar{s}_2(t; \boldsymbol{b}_2)$, can be expressed as

$$\begin{aligned} g(\tau; \boldsymbol{b}_1; \boldsymbol{b}_2) &= \int_{-\infty}^{\infty} \bar{s}_1(t; \boldsymbol{b}_1)\bar{s}_2^*(t-\tau; \boldsymbol{b}_2)dt \\ &= \int_{-\infty}^{\infty} \sum_{n=1}^{N-1} s_{1,n}\left(t-(n-1)T_s\right) \sum_{m=1}^{N-1} s_{2,m}^*\left(t-(m-1)T_s-\tau\right) dt \\ &= \sum_{n=1}^{N-1}\sum_{m=1}^{N-1} \int_{-\infty}^{\infty} s_{1,n}\left(t-(n-1)T_s\right) s_{2,m}^*\left(t-(m-1)T_s-\tau\right) dt \\ &= \sum_{n=1}^{N-1}\sum_{m=1}^{N-1} f_{n,m}^{1,2}(\tau), \end{aligned} \tag{6.5}$$

where $s_{i,n}\big(t-(n-1)T_s\big)$ denotes the nth sub-pulse of $\bar{s}_i(t; \boldsymbol{bi})$ with $n \in \{1, \ldots, N-1\}$ and $i \in \{1, 2\}$. Furthermore, $f^{1,2}n, m(\tau)$ represents the cross-correlation function between sub-pulses, defined as

$$f_{n,m}^{1,2}(\tau) = \int_{-\infty}^{\infty} s_{1,n}\left(t-(n-1)T_s\right) s_{2,m}^*\left(t-(m-1)T_s-\tau\right) dt. \tag{6.6}$$

By defining $t' = t - (n-1)T_s$ and denoting $b_{i,n}$ and $\alpha_{i,n}$ as the instantaneous frequency offset and slope of the nth sub-pulse of the ith waveform $\bar{s}_i(t; \boldsymbol{b}_i)$, Eq. (6.6) can be rewritten as

$$
\begin{aligned}
f_{n,m}^{1,2}(\tau) &= \int_{-\infty}^{\infty} s_{1,n}\left(t'\right) s_{2,m}^{*}\left(t' + (n-m)T_s - \tau\right) dt' \\
&= \exp\left(-j\pi\alpha_{2,m}\tau^2\right) \exp\left(j2\pi\left(b_{2,m} + \alpha_{2,m}(n-m)T_s\right)\tau\right) \\
&\quad \times \exp\left(-j2\pi b_{2,m}(n-m)T_s\right) \exp\left(-j\pi\alpha_{2,m}(n-m)^2T_s^2\right) \\
&quad \times \int_{-\infty}^{\infty} \mathrm{rect}\left(\frac{t' - T_s/2}{T_s}\right) \mathrm{rect}\left(\frac{t' + (n-m-1/2)T_s - \tau}{T_s}\right) \\
&\quad \times \exp\left(j\pi\left(\alpha_{1,n} - \alpha_{2,m}\right)t'^2\right) \\
&\quad \times \exp\left(j2\pi\left(\alpha_{2,m}\tau + b_{1,n} - b_{2,m} - \alpha_{2,m}(n-m)T_s\right)t'\right) dt' \\
&= \exp\left(-j\pi\alpha_{2,m}\tau^2\right) \exp\left(j2\pi\left(b_{1,n} - \Delta b_{n,m}^{1,2}\right)\tau\right) \Phi_{n,m}^{1,2} \\
&\quad \times \mathrm{rect}\left(\frac{\tau - (n-m)T_s}{2T_s}\right) \\
&\quad \times \int_{t_1'}^{t_2'} \exp\left(j\pi\Delta\alpha_{n,m}^{1,2} t'^2\right) \exp\left(j2\pi\left(\alpha_{2,m}\tau + \Delta b_{n,m}^{1,2}\right)t'\right) dt',
\end{aligned}
\tag{6.7}
$$

where

$$
\begin{aligned}
\Delta\alpha_{n,m}^{1,2} &= \alpha_{1,n} - \alpha_{2,m} \\
\Delta b_{n,m}^{1,2} &= b_{1,n} - b_{2,m} - \alpha_{2,m}(n-m)T_s \\
\Phi_{n,m}^{1,2} &= \exp\left(-j2\pi b_{2,m}(n-m)T_s\right) \exp\left(-j\pi\alpha_{2,m}(n-m)^2T_s^2\right)
\end{aligned}
\tag{6.8}
$$

and

$$
\begin{cases}
t_1' = 0 \quad \text{and} \quad t_2' = (1+m-n)T_s + \tau, \\
\qquad \text{if } (n-m-1)T_s < \tau \le (n-m)T_s \\
t_1' = \tau + (m-n)T_s \quad \text{and} \quad t_2' = T_s, \\
\qquad \text{if } (n-m)T_s \le \tau \le (1+n-m)T_s
\end{cases}.
\tag{6.9}
$$

To continue, various scenarios determined by the sign of $\Delta\alpha^{1,2}n, m$ need to be considered. When $\Delta\alpha^{1,2}n, m > 0$, Eq. (6.7) can be written as

$$
\begin{aligned}
f_{n,m}^{1,2}(\tau) &= \exp\left(-j\pi\left(\alpha_{2,m} + \frac{\left(\alpha_{2,m}\right)^2}{\Delta\alpha_{n,m}^{1,2}}\right)\tau^2\right) \\
&\quad \times \exp\left(j2\pi\left(b_{1,n} - \Delta b_{n,m}^{1,2} - \frac{\alpha_{2,m}\Delta b_{n,m}^{1,2}}{\Delta\alpha_{n,m}^{1,2}}\right)\tau\right) \Phi_{n,m}^{1,2} \\
&\quad \times \exp\left(-j\pi\frac{(\Delta b_{n,m}^{1,2})^2}{\Delta\alpha_{n,m}^{1,2}}\right) \mathrm{rect}\left(\frac{\tau - (n-m)T_s}{2T_s}\right) \\
&\quad \times \int_{t_1'}^{t_2'} \exp\left(j\pi\left(\sqrt{\Delta\alpha_{n,m}^{1,2}}t' + \frac{\alpha_{2,m}\tau + \Delta b_{n,m}^{1,2}}{\sqrt{\Delta\alpha_{n,m}^{1,2}}}\right)^2\right) dt'.
\end{aligned}
\tag{6.10}
$$

Hence, indicating with

$$\xi(t') = \sqrt{\Delta\alpha_{n,m}^{1,2}}t' + \frac{\alpha_{2,m} + \Delta b_{n,m}^{1,2}}{\sqrt{\Delta\alpha_{n,m}^{1,2}}} \tag{6.11}$$

and performing the change of variable

$$\xi = \sqrt{\Delta\alpha_{n,m}^{1,2}}t' + \frac{\alpha_{2,m} + \Delta b_{n,m}^{1,2}}{\sqrt{\Delta\alpha_{n,m}^{1,2}}}, \tag{6.12}$$

(6.10) can be further simplified as

$$\begin{aligned}
f_{n,m}^{1,2}(\tau) &= \exp\left(-j\pi\left(\alpha_{2,m} + \frac{(\alpha_{2,m})^2}{\Delta\alpha_{n,m}^{1,2}}\right)\tau^2\right) \\
&\times \exp\left(j2\pi\left(b_{1,n} - \Delta b_{n,m}^{1,2} - \frac{\alpha_{2,m}\Delta b_{n,m}^{1,2}}{\Delta\alpha_{n,m}^{1,2}}\right)\tau\right) \\
&\times \frac{\Phi_{n,m}^{1,2}\exp\left(-j\pi\frac{(\Delta b_{n,m}^{1,2})^2}{\Delta\alpha_{n,m}^{1,2}}\right)}{\sqrt{\Delta\alpha_{n,m}^{1,2}}}\operatorname{rect}\left(\frac{\tau-(n-m)T_s}{2T_s}\right)\int_{\xi(t_1')}^{\xi(t_2')}\exp\left(j\pi\xi^2\right)d\xi \\
&= \exp\left(-j\pi\left(\alpha_{2,m} + \frac{(\alpha_{2,m})^2}{\Delta\alpha_{n,m}^{1,2}}\right)\tau^2\right) \\
&\times \exp\left(j2\pi\left(b_{1,n} - \Delta b_{n,m}^{1,2} - \frac{\alpha_{2,m}\Delta b_{n,m}^{1,2}}{\Delta\alpha_{n,m}^{1,2}}\right)\tau\right)\frac{\Phi_{n,m}^{1,2}\exp\left(-j\pi\frac{(\Delta b_{n,m}^{1,2})^2}{\Delta\alpha_{n,m}^{1,2}}\right)}{\sqrt{\Delta\alpha_{n,m}^{1,2}}} \\
&\times \operatorname{rect}\left(\frac{\tau-(n-m)T_s}{2T_s}\right)\left(R\left(\xi\left(t_2'\right)\right) + jI\left(\xi\left(t_2'\right)\right) - R\left(\xi\left(t_1'\right)\right) - jI\left(\xi\left(t_1'\right)\right)\right).
\end{aligned} \tag{6.13}$$

In Eq. (6.13), $R(\xi)$ and $I(\xi)$ denote the Fresnel integrals [29], defined as

$$\begin{aligned}
R(\xi) &= \int_0^\xi \cos\left(\pi\xi^2\right)d\xi \\
I(\xi) &= \int_0^\xi \sin\left(\pi\xi^2\right)d\xi
\end{aligned}. \tag{6.14}$$

When $\Delta\alpha_{n,m}^{1,2} < 0$, Eq. (6.7) can be reformulated as

$$\begin{aligned} f_{n,m}^{1,2}(\tau) = & \exp\left(-j\pi\left(\alpha_{2,m} + \frac{(\alpha_{2,m})^2}{\Delta\alpha_{n,m}^{1,2}}\right)\tau^2\right) \\ & \times \exp\left(j2\pi\left(b_{1,n} - \Delta b_{n,m}^{1,2} - \frac{\alpha_{2,m}\Delta b_{n,m}^{1,2}}{\Delta\alpha_{n,m}^{1,2}}\right)\tau\right)\Phi_{n,m}^{1,2} \\ & \times \exp\left(-j\pi\frac{(\Delta b_{n,m}^{1,2})^2}{\Delta\alpha_{n,m}^{1,2}}\right)\mathrm{rect}\left(\frac{\tau-(n-m)T_s}{2T_s}\right) \\ & \times \int_{t_1'}^{t_2'} \exp\left(-j\pi\left(\sqrt{-\Delta\alpha_{n,m}^{1,2}}t' - \frac{\alpha_{2,m}\tau + \Delta b_{n,m}^{1,2}}{\sqrt{-\Delta\alpha_{n,m}^{1,2}}}\right)^2\right)dt'. \end{aligned} \tag{6.15}$$

Thus, letting

$$\xi'(t') = \sqrt{-\Delta\alpha_{n,m}^{1,2}}t' - \frac{\alpha_{2,m} + \Delta b_{n,m}^{1,2}}{\sqrt{-\Delta\alpha_{n,m}^{1,2}}} \tag{6.16}$$

and performing the change of variable

$$\xi' = \sqrt{-\Delta\alpha_{nm}^{1,2}}t' - \frac{\alpha_{2,m} + \Delta b_{n,m}^{1,2}}{\sqrt{-\Delta\alpha_{n,m}^{1,2}}}, \tag{6.17}$$

(6.15) becomes

$$\begin{aligned} f_{n,m}^{1,2}(\tau) = & \exp\left(-j\pi\left(\alpha_{2,m} + \frac{(\alpha_{2,m})^2}{\Delta\alpha_{n,m}^{1,2}}\right)\tau^2\right) \\ & \times \exp\left(j2\pi\left(b_{1,n} - \Delta b_{n,m}^{1,2} - \frac{\alpha_{2,m}\Delta b_{n,m}^{1,2}}{\sqrt{\Delta\alpha_{n,m}^{1,2}}}\right)\tau\right) \\ & \times \frac{\Phi_{n,m}^{1,2}\exp\left(-j\pi\frac{(\Delta b_{n,m}^{1,2})^2}{\Delta\alpha_{n,m}^{1,2}}\right)}{\sqrt{-\Delta\alpha_{n,m}^{1,2}}}\mathrm{rect}\left(\frac{\tau-(n-m)T_s}{2T_s}\right) \\ & \times \left(R\left(\xi'\left(t_2'\right)\right) - jI\left(\xi'\left(t_2'\right)\right) - R\left(\xi'\left(t_1'\right)\right) + jI\left(\xi'\left(t_1'\right)\right)\right). \end{aligned} \tag{6.18}$$

Finally, for $\Delta\alpha^{1,2}_{n,m} = 0$, Eq. (6.7) can be rewritten as (see Appendix 6.6)

$$\begin{aligned} f^{1,2}_{n,m}(\tau) &= \exp\left(-j\pi\alpha_{1,n}\tau^2\right)\exp\left(j2\pi\left(b_{1,n} - \Delta b^{1,2}_{n,m}\right)\tau\right) \\ &\times \Phi^{1,2}_{n,m}\exp\left(j\pi\left(\alpha_{1,n}\tau + \Delta b^{1,2}_{n,m}\right)(T_s + (m-n)T_s + \tau)\right) \\ &\times (T_s - |\tau + (m-n)T_s|)\,\mathrm{rect}\left(\frac{\tau - (n-m)T_s}{2T_s}\right) \\ &\times \mathrm{sinc}\left(\left(\alpha_{1,n}\tau + \Delta b^{1,2}_{n,m}\right)(T_s - |\tau + (m-n)T_s|)\right). \end{aligned} \tag{6.19}$$

Using the same approach as in Eq. (6.5), the auto-correlation function of $\bar{s}_i(t; \boldsymbol{b}_i)$ can be expressed as

$$h(\tau; \boldsymbol{b}_i) = \sum_{n=1}^{N-1}\sum_{m=1}^{N-1} f^{i,i}_{n,m}(\tau) = \sum_{n=1}^{N-1} f^{i,i}_{n,n}(\tau) + \sum_{n=1}^{N-1}\sum_{m=1,m\neq n}^{N-1} f^{i,i}_{n,m}(\tau), \tag{6.20}$$

where $f^{i,i}_{n,n}(\tau)$ denotes the auto-correlation of the nth sub-pulse of $\bar{s}i(t; \boldsymbol{b}i)$ (recall that $f^{i,i}_{n,m}(\tau)$ represents the cross-correlation between the nth and mth sub-pulses). This expression is derived from Eq. (6.19) by setting $n = m$, $\Delta b^{1,2}_{n,m} = 0$, and $\Phi^{1,2}_{n,m} = 1$, i.e.,

$$\begin{aligned} f^{i,i}_{n,n}(\tau) &= (T_s - |\tau|)\exp\left(j\pi(\alpha_{i,n}T_s + 2b_{i,n})\tau\right) \\ &\mathrm{rect}\left(\frac{\tau}{2T_s}\right)\mathrm{sinc}\left(\alpha_{i,n}\tau\,(T_s - |\tau|)\right). \end{aligned} \tag{6.21}$$

6.2.3 *Optimization Method*

The PSLR and ISLR are commonly used indicators for evaluating waveform imaging performance. Their formal definitions are given as

$$f_{\mathrm{PSLR}}(\boldsymbol{b}) = \max_{\tau}\left|\frac{h(\tau; \boldsymbol{b})}{h(0; \boldsymbol{b})}\right|, \tau \in [\tau_m, T], \tag{6.22}$$

$$f_{\mathrm{ISLR}}(\boldsymbol{b}) = \frac{\int_{\tau_m}^{T} |h(\tau; \boldsymbol{b})|^2 d\tau}{\int_0^{\tau_m} |h(\tau; \boldsymbol{b})|^2 d\tau}, \tag{6.23}$$

where the interval $[-\tau_m, \tau_m]$ represents the main-lobe temporal extent, with τ_m generally depending on $\boldsymbol{b}$. Additionally, the CCE metric, which quantifies the degree of orthogonality between waveforms, is defined as

$$f_{\mathrm{CCE}}(\boldsymbol{b}_1; \boldsymbol{b}_2) = \int_{-T}^{T} |g(\tau; \boldsymbol{b}_1; \boldsymbol{b}_2)|^2 d\tau. \tag{6.24}$$

In general, a PSLR value lower than −20 dB is desired. When time-varying system imperfections are taken into account, an ideal waveform typically achieves a PSLR of around −35 dB [26]. In what follows, we address the design of NLFM signals characterized by piecewise linear frequency trajectories, where both CCE and PSLR serve as evaluation metrics. The problem can be expressed as the following constrained optimization formulation:

$$P(\boldsymbol{b}_1; \boldsymbol{b}_2) \begin{cases} \min\limits_{\boldsymbol{b}_1; \boldsymbol{b}_2} f_{\mathrm{CCE}}(\boldsymbol{b}_1; \boldsymbol{b}_2) \\ \text{s.t. } f_{\mathrm{PSLR}}(\boldsymbol{b}_1) \le a_1, \; f_{\mathrm{PSLR}}(\boldsymbol{b}_2) \le a_2 \\ \boldsymbol{b}_1, \boldsymbol{b}_2 \in \Omega \end{cases}, \tag{6.25}$$

where a_1 and a_2 denote the design thresholds for PSLRs, and Ω defines the feasible set for the vector $\boldsymbol{b}$, i.e.,

$$\Omega = \left\{ \boldsymbol{b} \in \mathbb{R}^N : b_1 = -\frac{B}{2}, b_N = \frac{B}{2}, -\frac{B}{2} \le \boldsymbol{b} \le \frac{B}{2} \right\}. \tag{6.26}$$

To obtain an approximate solution to the non-convex optimization problem $P(\boldsymbol{b}_1; \boldsymbol{b}_2)$, the Coordinate Descent (CD) algorithm is employed. The overall procedure for generating the quasi-orthogonal NLFM waveforms is summarized in Table 6.1, where $\boldsymbol{b}_1^{(l)}$ and $\boldsymbol{b}_2^{(l)}$ represent the updated vectors at iteration l. In addition,

Table 6.1 Generation of quasi-orthogonal NLFM waveforms

Input: Given the signal bandwidth B and pulse duration T, together with an admissible initial frequency-control vector $\boldsymbol{b}_{\mathrm{Initial}}$ and the required minimum enhancement level δ;
Output: The Coordinate Descent (CD) solutions for the first ($\boldsymbol{b}_1^{\mathrm{CD}}$) and second ($\boldsymbol{b}_2^{\mathrm{CD}}$) signals;   **(1) Initialization.**   • Initialize $l = 0$ and set the frequency control vectors for the first and second signals as $\boldsymbol{b}_1^{(l)} = \boldsymbol{b}_{\mathrm{Initial}}$ and $\boldsymbol{b}_2^{(l)} = -\boldsymbol{b}_{\mathrm{Initial}}$.   **(2) Improvement.**   • With $\boldsymbol{b}_1^{(l)}$ fixed, optimize $\boldsymbol{b}_2$ by solving $P\left(\boldsymbol{b}_1^{(l)}; \boldsymbol{b}_2\right)$ using the initial value $\boldsymbol{b}_2 = -\boldsymbol{b}_1^{(l)}$ to obtain $\boldsymbol{b}_2^{(l+1)}$;   • With $\boldsymbol{b}_2^{(l+1)}$ fixed, optimize $\boldsymbol{b}_1$ by solving $P\left(\boldsymbol{b}_1; \boldsymbol{b}_2^{(l+1)}\right)$ using the initial value $\boldsymbol{b}_1 = -\boldsymbol{b}_2^{(l+1)}$ to obtain $\boldsymbol{b}_1^{(l+1)}$;   **(3) Stopping Criterion.**   • If $\left| f_{\mathrm{CCE}}\left(\boldsymbol{b}_1^{(l+1)}; \boldsymbol{b}_2^{(l+1)}\right) - f_{\mathrm{CCE}}\left(\boldsymbol{b}_1^{(l)}; \boldsymbol{b}_2^{(l)}\right) \right| < \delta$, the iteration stops; otherwise, set $l = l + 1$ and return to step (2);   **(4) Output.**   • Set $\boldsymbol{b}_1^{\mathrm{CD}} = \boldsymbol{b}_1^{(l+1)}$ and $\boldsymbol{b}_2^{\mathrm{CD}} = \boldsymbol{b}_2^{(l+1)}$.

the search scheme proposed in [30] (implemented in ©MATLAB as "*GlobalSearch*") is applied to solve the subproblems $P\left(\boldsymbol{b}_1^{(l)}; \boldsymbol{b}_2\right)$ and $P\left(\boldsymbol{b}_1; \boldsymbol{b}_2^{(l+1)}\right)$.

6.3 Conventional Quad-Pol SAR

This section examines how quasi-orthogonal NLFM waveforms are combined with APC within a conventional quad-pol SAR configuration and presents the corresponding formulations used to evaluate AASR and RASR performance.

6.3.1 *Ambiguity Performance Using Quasi-Orthogonal NLFM Waveforms*

In a conventional quad-pol configuration, H- and V-polarized pulses are transmitted in an alternating manner, and echoes are recorded simultaneously on both polarization channels. The corresponding received signals expressed in the range-frequency (f_t) and azimuth-time (η) domains are provided in [12].

$$
\begin{aligned}
\begin{pmatrix} \hat{S}_{\mathrm{HH}}(f_t,\eta) & \hat{S}_{\mathrm{HV}}(f_t,\eta) \\ \hat{S}_{\mathrm{VH}}(f_t,\eta) & \hat{S}_{\mathrm{VV}}(f_t,\eta) \end{pmatrix} &= \begin{pmatrix} \tilde{S}_{\mathrm{HH}}(f_t,\eta) & \tilde{S}_{\mathrm{HV}}(f_t,\eta) \\ \tilde{S}_{\mathrm{VH}}(f_t,\eta) & \tilde{S}_{\mathrm{VV}}(f_t,\eta) \end{pmatrix} \\
&\quad + \sum_{k=\text{odd}} \begin{pmatrix} \check{S}_{\mathrm{HV}_k}(f_t,\eta) & \check{S}_{\mathrm{HH}_k}(f_t,\eta) \\ \check{S}_{\mathrm{VV}_k}(f_t,\eta) & \check{S}_{\mathrm{VH}_k}(f_t,\eta) \end{pmatrix} \\
&\quad + \sum_{k=\text{even}} \begin{pmatrix} \check{S}_{\mathrm{HH}_k}(f_t,\eta) & \check{S}_{\mathrm{HV}_k}(f_t,\eta) \\ \check{S}_{\mathrm{VH}_k}(f_t,\eta) & \check{S}_{\mathrm{VV}_k}(f_t,\eta) \end{pmatrix},
\end{aligned} \tag{6.27}
$$

where $\hat{S}pq\,(f_t,\eta)$ $(p,q \in \{\mathrm{H},\mathrm{V}\})$ denotes the measured echo for the pq polarization channel, where p and q specify the receive and transmit polarizations, respectively. The quantities $\tilde{S}pq\,(f_t,\eta)$ and $\check{S}pq_k\,(f_t,\eta)$ correspond to the desired return and the components arising from range ambiguities, with $k \in \mathbb{N},\, ,k \neq 0$ identifying the kth ambiguous contribution. The RASR is defined as the power ratio between the ambiguous returns and the desired signal [31–33]. Incorporating the Doppler-spectrum weighting function $W(f_\eta)$, the post–range-compression RASR for a single-polarization SAR operating without waveform diversity is given in [15].

$$\text{RASR} \cong \frac{\sum_{k=1}^{N_A} \frac{\sigma^0(\phi_k) \int_{-Ba/2}^{Ba/2} G_r^2(\theta_k, f_\eta) W^2(f_\eta) df_\eta}{R_k^3 \sin(\phi_k)}}{\frac{\sigma^0(\phi_0) \int_{-Ba/2}^{Ba/2} G_r^2(\theta_0, f_\eta) W^2(f_\eta) df_\eta}{R_0^3 \sin(\phi_0)}}. \tag{6.28}$$

where $\sigma^0(\phi)$ represents the backscatter reflectivity as a function of the incidence angle ϕ, and $G_r^2(\theta, f_\eta)$ denotes the two-way antenna power pattern in elevation, depending on the look angle θ and azimuth frequency f_η. The Doppler bandwidth after processing is B_a. The look and incidence angles of the kth range ambiguity (and the desired signal) are given by θ_k and ϕ_k (θ_0 and ϕ_0), respectively, and N_A specifies the total number of range ambiguities. Additionally, R_0 and R_k indicate the ranges of the desired signal and the kth range-ambiguous signal, respectively, i.e.,

$$R_k = R_0 + k\frac{\text{PRI}}{2}c, \tag{6.29}$$

where PRI represents the pulse repetition interval and c denotes the speed of light.

This study differs from [12, 15] by explicitly considering the influence of the transmitted waveforms, employing separate pulses $s_1(t)$ and $s_2(t)$ for H- and V-polarized transmissions. The timing diagram corresponding to this setup is presented in Fig. 6.2, while Table 6.2 summarizes the range compression procedure for each channel, with $S_i(f_t)$ denoting the Fourier transform of $s_i(t)$. Specifically, the signal after range compression in the range frequency and azimuth time domains is expressed by Eq. (6.30).

$$\begin{aligned}
&\begin{pmatrix} \hat{S}_{\text{HH}}^1(f_t,\eta) & \hat{S}_{\text{HV}}^2(f_t,\eta) \\ \hat{S}_{\text{VH}}^1(f_t,\eta) & \hat{S}_{\text{VV}}^2(f_t,\eta) \end{pmatrix} \odot \begin{pmatrix} S_1^*(f_t) & S_2^*(f_t) \\ S_1^*(f_t) & S_2^*(f_t) \end{pmatrix} = \\
&\left\{ \begin{pmatrix} \tilde{S}_{\text{HH}}^1(f_t,\eta) & \tilde{S}_{\text{HV}}^2(f_t,\eta) \\ \tilde{S}_{\text{VH}}^1(f_t,\eta) & \tilde{S}_{\text{VV}}^2(f_t,\eta) \end{pmatrix} + \sum_{k=\text{odd}} \begin{pmatrix} \check{S}_{\text{HV}_k}^2(f_t,\eta) & \check{S}_{\text{HH}_k}^1(f_t,\eta) \\ \check{S}_{\text{VV}_k}^2(f_t,\eta) & \check{S}_{\text{VH}_k}^1(f_t,\eta) \end{pmatrix} \right. \\
&\left. + \sum_{k=\text{even}} \begin{pmatrix} \check{S}_{\text{HH}_k}^1(f_t,\eta) & \check{S}_{\text{HV}_k}^2(f_t,\eta) \\ \check{S}_{\text{VH}_k}^1(f_t,\eta) & \check{S}_{\text{VV}_k}^2(f_t,\eta) \end{pmatrix} \right\} \odot \begin{pmatrix} S_1^*(f_t) & S_2^*(f_t) \\ S_1^*(f_t) & S_2^*(f_t) \end{pmatrix} \\
&= \begin{pmatrix} \tilde{S}_{\text{HH}}^{1,1}(f_t,\eta) & \tilde{S}_{\text{HV}}^{2,2}(f_t,\eta) \\ \tilde{S}_{\text{VH}}^{1,1}(f_t,\eta) & \tilde{S}_{\text{VV}}^{2,2}(f_t,\eta) \end{pmatrix} + \sum_{k=\text{odd}} \begin{pmatrix} \check{S}_{\text{HV}_k}^{2,1}(f_t,\eta) & \check{S}_{\text{HH}_k}^{1,2}(f_t,\eta) \\ \check{S}_{\text{VV}_k}^{2,1}(f_t,\eta) & \check{S}_{\text{VH}_k}^{1,2}(f_t,\eta) \end{pmatrix} \\
&+ \sum_{k=\text{even}} \begin{pmatrix} \check{S}_{\text{HH}_k}^{1,1}(f_t,\eta) & \check{S}_{\text{HV}_k}^{2,2}(f_t,\eta) \\ \check{S}_{\text{VH}_k}^{1,1}(f_t,\eta) & \check{S}_{\text{VV}_k}^{2,2}(f_t,\eta) \end{pmatrix},
\end{aligned} \tag{6.30}$$

where $\hat{S}_{pq}^i(f_t,\eta)$ $(p, q \in \{\text{H}, \text{V}\}, i \in \{1, 2\})$ denotes the signal observed at the pq polarimetric channel produced by the waveform $S_i(f_t)$. The quantities $\tilde{S}_{pq}^{i,j}(f_t,\eta)$ and $\check{S}_{pq_k}^{i,j}(f_t,\eta)$ $(p, q \in \{\text{H}, \text{V}\}, i, j \in \{1, 2\})$ represent the outputs of the range matched

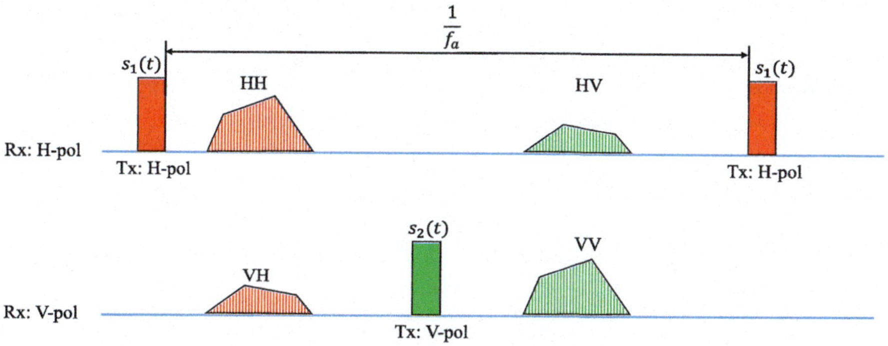

Fig. 6.2 For a conventional quad-pol SAR system, the timing diagram illustrates that transmitted pulses with the same polarization have a pulse repetition frequency denoted by f_a

Table 6.2 Processing of desired and range-ambiguous signals via range compression

Channel	Desired signal	Odd ambiguities	Even ambiguities
HH	$S_1(f_t)S_1^*(f_t)$	$S_2(f_t)S_1^*(f_t)$	$S_1(f_t)S_1^*(f_t)$
HV	$S_2(f_t)S_2^*(f_t)$	$S_1(f_t)S_2^*(f_t)$	$S_2(f_t)S_2^*(f_t)$
VH	$S_1(f_t)S_1^*(f_t)$	$S_2(f_t)S_1^*(f_t)$	$S_1(f_t)S_1^*(f_t)$
VV	$S_2(f_t)S_2^*(f_t)$	$S_1(f_t)S_2^*(f_t)$	$S_2(f_t)S_2^*(f_t)$

filter $S_j^*(f_t)$ when applied to the desired return $\tilde{S}^i_{pq}(f_t, \eta)$ (the desired return produced by $S_i(f_t)$) and the kth range-ambiguous component $\check{S}^i_{pq_k}(f_t, \eta)$ generated by $S_i(f_t)$, respectively.

By taking into account the specific values of $\sigma_0(\phi)$ for each channel's contribution, the RASR can be written as in Eq. (6.31), under the assumption that the antenna patterns are identical across all polarizations. Cross-polarized returns $(\sigma^0_{\mathrm{HV}}(\phi_k), \sigma^0_{\mathrm{VH}}(\phi_k))$ are generally 6–10 dB weaker than the corresponding co-polarized signals $(\sigma^0_{\mathrm{HH}}(\phi_k), \sigma^0_{\mathrm{VV}}(\phi_k))$. Moreover, odd-numbered range ambiguities observed in the HV and VH channels originate from HH and VV contributions, which can result in relatively strong ambiguities in the cross-polarized channels. In general, employing diverse quasi-orthogonal NLFM waveforms reduces the CCE, i.e., $\int_{-T}^{T} |g(\tau, \boldsymbol{b}_1, \boldsymbol{b}_2)|^2 d\tau$, thereby significantly lowering the energy associated with the range ambiguities.

$$
\begin{aligned}
\mathrm{RASR_{HH}} &\cong \frac{\sum_{k=\mathrm{odd}}^{N_A} \dfrac{\sigma_{\mathrm{HV}}^0(\phi_k) \int_{-\tau_m}^{\tau_m} |g(\tau;\boldsymbol{b}_2;\boldsymbol{b}_1)|^2 d\tau \int_{-Ba/2}^{Ba/2} G_r^2(\theta_k, f_\eta) W^2(f_\eta) df_\eta}{R_k^3 \sin(\phi_k)}}{\dfrac{\sigma_{\mathrm{HH}}^0(\phi_0) \int_{-\tau_m}^{\tau_m} |h(\tau;\boldsymbol{b}_1)|^2 d\tau \int_{-Ba/2}^{Ba/2} G_r^2(\theta_0, f_\eta) W^2(f_\eta) df_\eta}{R_0^3 \sin(\phi_0)}} \\
&+ \frac{\sum_{k=\mathrm{even},\neq 0}^{N_A} \dfrac{\sigma_{\mathrm{HH}}^0(\phi_k) \int_{-Ba/2}^{Ba/2} G_r^2(\theta_k, f_\eta) W^2(f_\eta) df_\eta}{R_k^3 \sin(\phi_k)}}{\dfrac{\sigma_{\mathrm{HH}}^0(\phi_0) \int_{-Ba/2}^{Ba/2} G_r^2(\theta_0, f_\eta) W^2(f_\eta) df_\eta}{R_0^3 \sin(\phi_0)}} \\
\mathrm{RASR_{HV}} &\cong \frac{\sum_{k=\mathrm{odd}}^{N_A} \dfrac{\sigma_{\mathrm{HH}}^0(\phi_k) \int_{-\tau_m}^{\tau_m} |g(\tau;\boldsymbol{b}_1;\boldsymbol{b}_2)|^2 d\tau \int_{-Ba/2}^{Ba/2} G_r^2(\theta_k, f_\eta) W^2(f_\eta) df_\eta}{R_k^3 \sin(\phi_k)}}{\dfrac{\sigma_{\mathrm{HV}}^0(\phi_0) \int_{-\tau_m}^{\tau_m} |h(\tau;\boldsymbol{b}_2)|^2 d\tau \int_{-Ba/2}^{Ba/2} G_r^2(\theta_0, f_\eta) W^2(f_\eta) df_\eta}{R_0^3 \sin(\phi_0)}} \\
&+ \frac{\sum_{k=\mathrm{even},\neq 0}^{N_A} \dfrac{\sigma_{\mathrm{HV}}^0(\phi_k) \int_{-Ba/2}^{Ba/2} G_r^2(\theta_k, f_\eta) W^2(f_\eta) df_\eta}{R_k^3 \sin(\phi_k)}}{\dfrac{\sigma_{\mathrm{HV}}^0(\phi_0) \int_{-Ba/2}^{Ba/2} G_r^2(\theta_0, f_\eta) W^2(f_\eta) df_\eta}{R_0^3 \sin(\phi_0)}} \\
\mathrm{RASR_{VH}} &\cong \frac{\sum_{k=\mathrm{odd}}^{N_A} \dfrac{\sigma_{\mathrm{VV}}^0(\phi_k) \int_{-\tau_m}^{\tau_m} |g(\tau;\boldsymbol{b}_2;\boldsymbol{b}_1)|^2 d\tau \int_{-Ba/2}^{Ba/2} G_r^2(\theta_k, f_\eta) W^2(f_\eta) df_\eta}{R_k^3 \sin(\phi_k)}}{\dfrac{\sigma_{\mathrm{VH}}^0(\phi_0) \int_{-\tau_m}^{\tau_m} |h(\tau;\boldsymbol{b}_1)|^2 d\tau \int_{-Ba/2}^{Ba/2} G_r^2(\theta_0, f_\eta) W^2(f_\eta) df_\eta}{R_0^3 \sin(\phi_0)}} \\
&+ \frac{\sum_{k=\mathrm{even},\neq 0}^{N_A} \dfrac{\sigma_{\mathrm{VH}}^0(\phi_k) \int_{-Ba/2}^{Ba/2} G_r^2(\theta_k, f_\eta) W^2(f_\eta) df_\eta}{R_k^3 \sin(\phi_k)}}{\dfrac{\sigma_{\mathrm{VH}}^0(\phi_0) \int_{-Ba/2}^{Ba/2} G_r^2(\theta_0, f_\eta) W^2(f_\eta) df_\eta}{R_0^3 \sin(\phi_0)}} \\
\mathrm{RASR_{VV}} &\cong \frac{\sum_{k=\mathrm{odd}}^{N_A} \dfrac{\sigma_{\mathrm{VH}}^0(\phi_k) \int_{-\tau_m}^{\tau_m} |g(\tau;\boldsymbol{b}_1;\boldsymbol{b}_2)|^2 d\tau \int_{-Ba/2}^{Ba/2} G_r^2(\theta_k, f_\eta) W^2(f_\eta) df_\eta}{R_k^3 \sin(\phi_k)}}{\dfrac{\sigma_{\mathrm{VV}}^0(\phi_0) \int_{-\tau_m}^{\tau_m} |h(\tau;\boldsymbol{b}_2)|^2 d\tau \int_{-Ba/2}^{Ba/2} G_r^2(\theta_0, f_\eta) W^2(f_\eta) df_\eta}{R_0^3 \sin(\phi_0)}} \\
&+ \frac{\sum_{k=\mathrm{even},\neq 0}^{N_A} \dfrac{\sigma_{\mathrm{VV}}^0(\phi_k) \int_{-Ba/2}^{Ba/2} G_r^2(\theta_k, f_\eta) W^2(f_\eta) df_\eta}{R_k^3 \sin(\phi_k)}}{\dfrac{\sigma_{\mathrm{VV}}^0(\phi_0) \int_{-Ba/2}^{Ba/2} G_r^2(\theta_0, f_\eta) W^2(f_\eta) df_\eta}{R_0^3 \sin(\phi_0)}}
\end{aligned}
\tag{6.31}
$$

For a uniformly scattering scene, azimuth ambiguity-to-signal ratio (AASR) is defined as the proportion of azimuth-ambiguous signal power relative to the main signal [34]. In a conventional quad-pol SAR configuration, the AASR corresponding to each polarimetric channel after range-matched filtering can be calculated as described in [15].

$$\text{AASR} \cong \frac{\sum_{k=-\infty,\neq 0}^{\infty} \int_{-B_a/2}^{B_a/2} G_a^2(f_\eta + kf_a)W^2(f_\eta)df_\eta}{\int_{-B_a/2}^{B_a/2} G_a^2(f_\eta)W^2(f_\eta)df_\eta}, \tag{6.32}$$

where $G_a^2(f_\eta)$ denotes the two-way azimuth antenna power pattern, and f_a corresponds to the PRF of transmitted pulses with the same polarization. The formula remains valid even when different pulses are employed.

6.3.2 Extension Using a Combination of Quasi-Orthogonal NLFM Waveforms and APC

Table 6.3 presents an intertwined scheme combining quasi-orthogonal NLFM waveforms with the APC method, with the corresponding matched filter outputs summarized in Table 6.4. It can be observed that the periodic variations (within a given polarimetric channel) for the odd-numbered ambiguous signals corresponding to Tx-H and Tx-V are $\{S_1^*(f_t)S_2(f_t), S_1^*(f_t)S_2(f_t)e^{-j\pi}\}$ and $\{S_1(f_t)S_2^*(f_t), S_1(f_t)S_2^*(f_t)e^{j\pi}\}$, respectively. Consequently, the odd-numbered ambiguities are more effectively sup-

Table 6.3 Integration of quasi-orthogonal NLFM waveforms with the APC method

Tx polarization:	H	V	H	V
Transmitted signal:	$S_1(f_t)$	$S_2(f_t)$	$S_1(f_t)e^{j\pi}$	$S_2(f_t)$
Odd-numbered ambiguous signal:	$S_2(f_t)$	$S_1(f_t)$	$S_2(f_t)$	$S_1(f_t)e^{j\pi}$
Even-numbered ambiguous signal:	$S_1(f_t)e^{j\pi}$	$S_2(f_t)$	$S_1(f_t)$	$S_2(f_t)$
Range matched filter:	$S_1^*(f_t)$	$S_2^*(f_t)$	$S_1^*(f_t)e^{-j\pi}$	$S_2^*(f_t)$

Table 6.4 Frequency-domain matched filter output for desired and range-ambiguous signals

Tx polarization:	H	V	H	V
Desired signal	$\|S_1(f_t)\|^2$	$\|S_2(f_t)\|^2$	$\|S_1(f_t)\|^2$	$\|S_2(f_t)\|^2$
Odd-numbered ambiguous signal	$S_1^*(f_t)S_2(f_t)$	$S_1(f_t)S_2^*(f_t)$	$S_1^*(f_t)S_2(f_t)e^{-j\pi}$	$S_1(f_t)S_2^*(f_t)e^{j\pi}$
Even-numbered ambiguous signal	$\|S_1(f_t)\|^2e^{j\pi}$	$\|S_2(f_t)\|^2$	$\|S_1(f_t)\|^2e^{-j\pi}$	$\|S_2(f_t)\|^2$

pressed due to the combined effect of waveform diversity and APC. The resulting RASRs can be calculated according to Eq. (6.33).

$$
\begin{aligned}
\mathrm{RASR}_{\mathrm{HH}}^{\mathrm{APC}} &\cong \frac{\sum\limits_{k=\mathrm{even},\neq 0}^{N_A} \frac{\sigma_{\mathrm{HH}}^{0}(\phi_k)\int_{-Ba/2}^{Ba/2} G_r^2(\theta_k, f_\eta) W^2(f_\eta) df_\eta}{R_k^3 \sin(\phi_k)}}{\frac{\sigma_{\mathrm{HH}}^{0}(\phi_0)\int_{-Ba/2}^{Ba/2} G_r^2(\theta_0, f_\eta) W^2(f_\eta) df_\eta}{R_0^3 \sin(\phi_0)}} + \\
&\frac{\sum\limits_{k=\mathrm{odd}}^{N_A} \frac{\sigma_{\mathrm{HV}}^{0}(\phi_k)\int_{-\tau_m}^{\tau_m} |g(\tau;\boldsymbol{b}_2;\boldsymbol{b}_1)|^2 d\tau \int_{-Ba/2}^{Ba/2}\left[G_r^2\left(\theta_k, f_\eta+\frac{f_a}{2}\right)+G_r^2\left(\theta_k, f_\eta-\frac{f_a}{2}\right)\right] W^2(f_\eta) df_\eta}{\int_{-\tau_m}^{\tau_m} |h(\tau;\boldsymbol{b}_1)|^2 d\tau R_k^3 \sin(\phi_k)}}{\frac{\sigma_{\mathrm{HH}}^{0}(\phi_0)\int_{-Ba/2}^{Ba/2} G_r^2(\theta_0, f_\eta) W^2(f_\eta) df_\eta}{R_0^3 \sin(\phi_0)}} \\
\mathrm{RASR}_{\mathrm{HV}}^{\mathrm{APC}} &\cong \frac{\sum\limits_{k=\mathrm{even},\neq 0}^{N_A} \frac{\sigma_{\mathrm{HV}}^{0}(\phi_k)\int_{-Ba/2}^{Ba/2} G_r^2(\theta_k, f_\eta) W^2(f_\eta) df_\eta}{R_k^3 \sin(\phi_k)}}{\frac{\sigma_{\mathrm{HV}}^{0}(\phi_0)\int_{-Ba/2}^{Ba/2} G_r^2(\theta_0, f_\eta) W^2(f_\eta) df_\eta}{R_0^3 \sin(\phi_0)}} + \\
&\frac{\sum\limits_{k=\mathrm{odd}}^{N_A} \frac{\sigma_{\mathrm{HH}}^{0}(\phi_k)\int_{-\tau_m}^{\tau_m} |g(\tau;\boldsymbol{b}_1;\boldsymbol{b}_2)|^2 d\tau \int_{-Ba/2}^{Ba/2}\left[G_r^2\left(\theta_k, f_\eta+\frac{f_a}{2}\right)+G_r^2\left(\theta_k, f_\eta-\frac{f_a}{2}\right)\right] W^2(f_\eta) df_\eta}{\int_{-\tau_m}^{\tau_m} |h(\tau;\boldsymbol{b}_2)|^2 d\tau R_k^3 \sin(\phi_k)}}{\frac{\sigma_{\mathrm{HV}}^{0}(\phi_0)\int_{-Ba/2}^{Ba/2} G_r^2(\theta_0, f_\eta) W^2(f_\eta) df_\eta}{R_0^3 \sin(\phi_0)}} \\
\mathrm{RASR}_{\mathrm{VH}}^{\mathrm{APC}} &\cong \frac{\sum\limits_{k=\mathrm{even},\neq 0}^{N_A} \frac{\sigma_{\mathrm{VH}}^{0}(\phi_k)\int_{-Ba/2}^{Ba/2} G_r^2(\theta_k, f_\eta) W^2(f_\eta) df_\eta}{R_k^3 \sin(\phi_k)}}{\frac{\sigma_{\mathrm{VH}}^{0}(\phi_0)\int_{-Ba/2}^{Ba/2} G_r^2(\theta_0, f_\eta) W^2(f_\eta) df_\eta}{R_0^3 \sin(\phi_0)}} + \\
&\frac{\sum\limits_{k=\mathrm{odd}}^{N_A} \frac{\sigma_{\mathrm{VV}}^{0}(\phi_k)\int_{-\tau_m}^{\tau_m} |g(\tau;\boldsymbol{b}_2;\boldsymbol{b}_1)|^2 d\tau \int_{-Ba/2}^{Ba/2}\left[G_r^2\left(\theta_k, f_\eta+\frac{f_a}{2}\right)+G_r^2\left(\theta_k, f_\eta-\frac{f_a}{2}\right)\right] W^2(f_\eta) df_\eta}{\int_{-\tau_m}^{\tau_m} |h(\tau;\boldsymbol{b}_1)|^2 d\tau R_k^3 \sin(\phi_k)}}{\frac{\sigma_{\mathrm{VH}}^{0}(\phi_0)\int_{-Ba/2}^{Ba/2} G_r^2(\theta_0, f_\eta) W^2(f_\eta) df_\eta}{R_0^3 \sin(\phi_0)}} \\
\mathrm{RASR}_{\mathrm{VV}}^{\mathrm{APC}} &\cong \frac{\sum\limits_{k=\mathrm{even},\neq 0}^{N_A} \frac{\sigma_{\mathrm{VV}}^{0}(\phi_k)\int_{-Ba/2}^{Ba/2} G_r^2(\theta_k, f_\eta) W^2(f_\eta) df_\eta}{R_k^3 \sin(\phi_k)}}{\frac{\sigma_{\mathrm{VV}}^{0}(\phi_0)\int_{-Ba/2}^{Ba/2} G_r^2(\theta_0, f_\eta) W^2(f_\eta) df_\eta}{R_0^3 \sin(\phi_0)}} + \\
&\frac{\sum\limits_{k=\mathrm{odd}}^{N_A} \frac{\sigma_{\mathrm{VH}}^{0}(\phi_k)\int_{-\tau_m}^{\tau_m} |g(\tau;\boldsymbol{b}_1;\boldsymbol{b}_2)|^2 d\tau \int_{-Ba/2}^{Ba/2}\left[G_r^2\left(\theta_k, f_\eta+\frac{f_a}{2}\right)+G_r^2\left(\theta_k, f_\eta-\frac{f_a}{2}\right)\right] W^2(f_\eta) df_\eta}{\int_{-\tau_m}^{\tau_m} |h(\tau;\boldsymbol{b}_2)|^2 d\tau R_k^3 \sin(\phi_k)}}{\frac{\sigma_{\mathrm{VV}}^{0}(\phi_0)\int_{-Ba/2}^{Ba/2} G_r^2(\theta_0, f_\eta) W^2(f_\eta) df_\eta}{R_0^3 \sin(\phi_0)}}
\end{aligned}
\tag{6.33}
$$

6.4 Simulation and Performance Analysis

6.4.1 Waveform Performance Evaluation

For illustration, the LT-1 system parameters are adopted, featuring a bandwidth of 60 MHz and a pulse duration of 80 μs. The initial frequency-control vector, $\boldsymbol{b}_{\text{Initial}}$ ($N = 130$), is generated by sampling an NLFM waveform whose spectral shape closely matches a Taylor window with a maximum sidelobe level of -35 dB, as reported in [26]. The peak sidelobe ratio (PSLR) thresholds and convergence criterion are set to -35 dB and $\delta = 10^{-6}$, respectively.

Figure 6.3 shows the matched filter responses for quasi-orthogonal NLFM, up–down chirp, and OFDM-chirp waveforms [28].[2] For both the proposed NLFM and up–down chirp signals, the corresponding results can be derived by applying the theoretical expressions in Eqs. (6.5) and (6.20) to the vectors $\boldsymbol{b}_1$ and $\boldsymbol{b}_2$. Specifically, for the up–down chirp case, the vectors $\boldsymbol{b}_1$ and $\boldsymbol{b}_2$ are defined as

$$b_{1,i} = -\frac{B}{2} + \frac{B}{N-1}(i-1), i = 1, 2, \ldots, N, \tag{6.34}$$

and

$$b_{2,i} = \frac{B}{2} - \frac{B}{N-1}(i-1), i = 1, 2, \ldots, N, \tag{6.35}$$

For the proposed design, the vectors are generated using the implemented algorithm. The PSLR, ISLR, and CCE can then be calculated by applying the theoretical formulas in Eqs. (6.22), (6.23), and (6.24).

Figure 6.4 illustrates the ambiguity functions of the designed quasi-orthogonal NLFM waveforms. Key metrics for this waveform pair—including PSLR, ISLR, 3-dB main-lobe width (MW), and CCE—are summarized in Table 6.5. As observed in Fig. 6.3, the OFDM-chirp waveforms display grating lobes, whereas the quasi-orthogonal NLFM waveforms yield matched filter outputs (Fig. 6.3) and auto-ambiguity functions (Fig. 6.4) that are largely in agreement. Table 6.5 indicates that the CCE of the quasi-orthogonal NLFM signals is 3.47 dB lower than that of the up–down chirp waveforms. Furthermore, the PSLRs—already enforced during the design process—and the ISLRs of the synthesized quasi-orthogonal NLFM waveforms remain sufficiently low to satisfy SAR imaging requirements.

6.4.2 Point Target Simulation

This subsection reports simulation experiments using a point target to assess the ambiguity suppression performance of various Tx/Rx configurations. A standard

[2] All waveforms share a bandwidth of 60 MHz and a pulse duration of 80 μs.

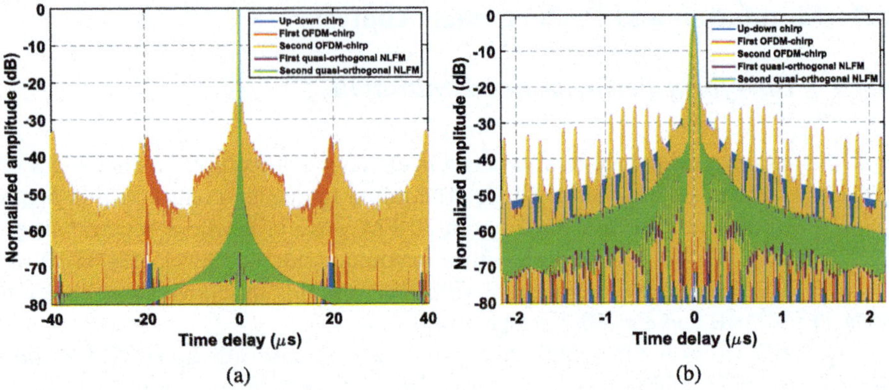

Fig. 6.3 Matched filter outputs (in dB) for various quasi-orthogonal waveforms. **a** Full view. **b** Zoomed view around the main lobe

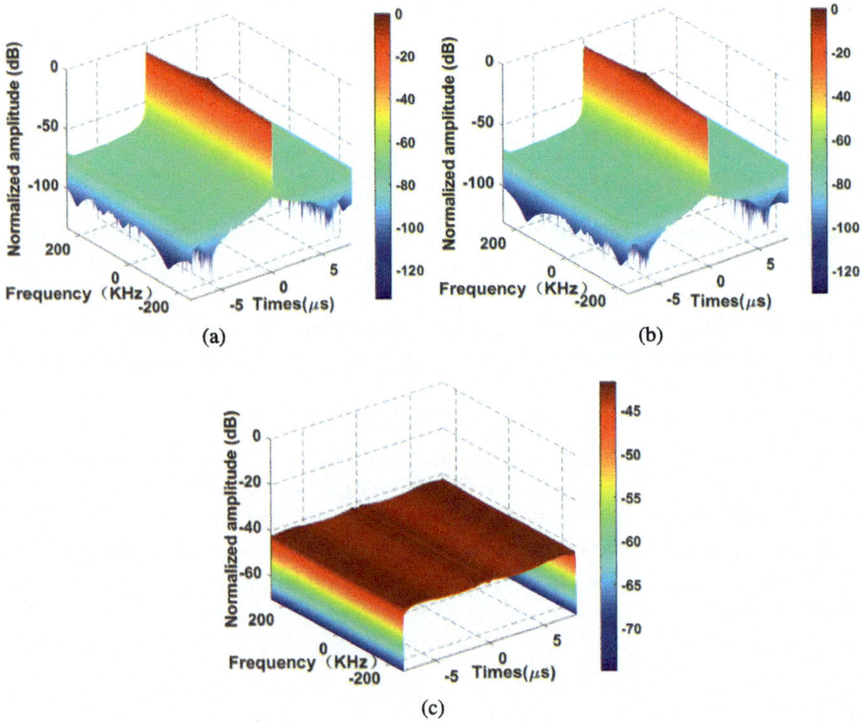

Fig. 6.4 Ambiguity functions (in dB) for the designed quasi-orthogonal NLFM waveforms are shown. Panels **a** and **b** display the auto-ambiguity functions of the first and second waveforms, respectively, whereas panel **c** illustrates the cross-ambiguity function between them

Table 6.5 Performance evaluation metrics for the various waveforms

Waveforms	PSLR (dB)	ISLR (dB)	3-dB MW (sample)	CCE[a] (dB)
Up-down chirp	−13.3	−9.7	1	0
Initial NLFM	−35	−26.6	1.33	0.91
First OFDM-chirp [28]	−13.26	−7.4	1	0.13
Second OFDM-chirp [28]	−13.29	−7.39	1	
First quasi-orthogonal NLFM	−35.7	−28	1.57	−3.47
Second quasi-orthogonal NLFM	−35.6	−28	1.56	

[a]The CCE relative to a chirp signal

Table 6.6 Key simulation parameters of LT-1 in quad-pol mode

Parameters	Value
Orbit height	600 Km
Look angle access range	25–27 deg
SAR pulse width	80 μs
Bandwidth	60 MHz
PRF ($2f_a$)	3320 Hz
Antenna type	Planar
Two-way antenna pattern	$\text{sinc}^2(\cdot)$
Processed Doppler bandwidth B_a	1380 HZ
Azimuth processing window	Generalized Hamming ($\alpha = 0.6$)
Backscatter model	Soil and Rock [35]

quad-pol SAR system, with parameters listed in Table 6.6, is employed, and imaging is performed using the Range Doppler Algorithm (RDA) [31].

Initially, a single-polarization SAR system is considered, utilizing the synthesized signal $s_1(t)$. The 2-D impulse response function (IRF) [31] for a point target, positioned as shown in Fig. 6.5 within the non-ambiguous region, is depicted in Fig. 6.6. The figure clearly demonstrates that the point target is accurately focused.

To evaluate the interference at the first near-range ambiguity from a point target located at midrange, Fig. 6.7 presents the 2-D IRFs for different transmission schemes within the first near-range ambiguous region of the image (HV channel in a conventional quad-pol system). In these simulations, the raw signals are assumed to have equal energy, and the azimuth signal is band-limited,[3] while factors such as antenna gain, backscatter reflectivity, incidence angle, and slant range are neglected. Table 6.7 lists the total measured ambiguity energies relative to a chirp waveform. When APC is applied, Doppler shifts and azimuth filtering reduce the total range ambiguity energy by approximately 0.81 dB compared to the chirp. Using quasi-orthogonal NLFM waveforms spreads and defocuses the ambiguous signals along

[3] The azimuth antenna pattern is not considered.

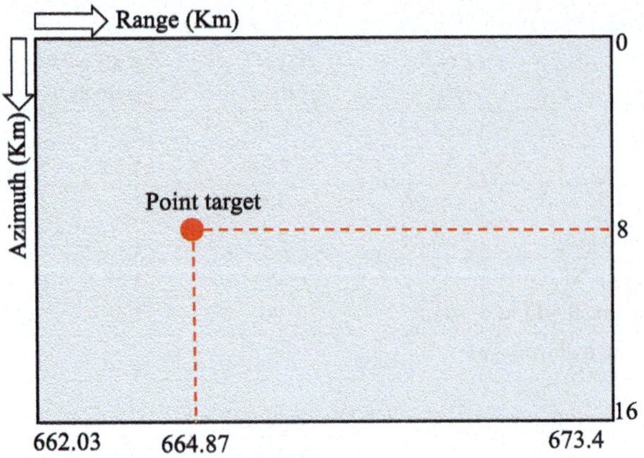

Fig. 6.5 Position of a point target within the non-ambiguous imaging region

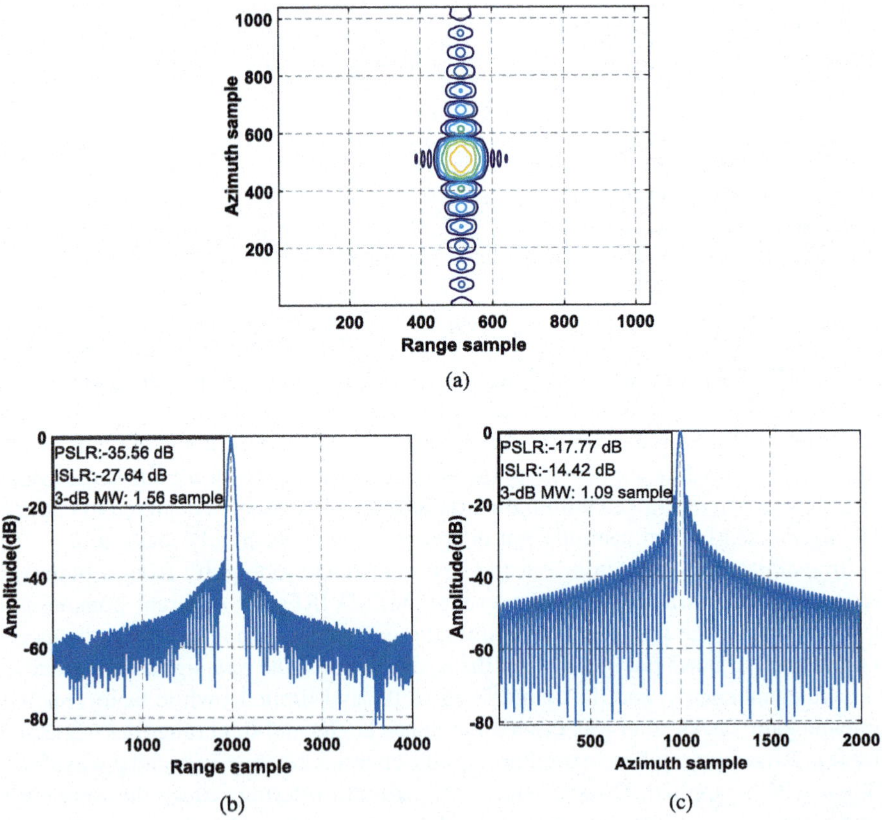

Fig. 6.6 2-D IRF of the point target depicted in Fig. 6.5, corresponding to the non-ambiguous portion of the image. The elevation antenna pattern is ignored, while the azimuth pattern is modeled as a $\text{sinc}^2(\cdot)$ function. Panel **a** shows the contour plot in dB, panel **b** presents the range profile in dB, and panel **c** depicts the azimuth profile in dB

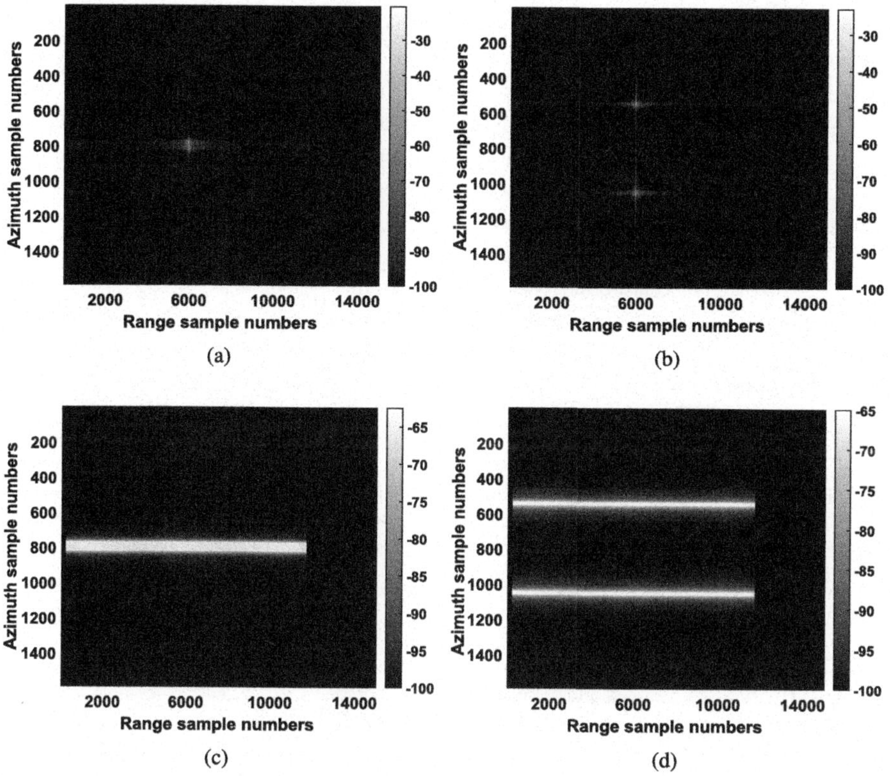

Fig. 6.7 2-D IRFs (in dB) of the first near-range ambiguous region of the image (HV polarization channel in a conventional quad-pol mode) are shown for: **a** up chirp waveform, **b** APC, **c** quasi-orthogonal NLFM waveforms, and **d** the combination of quasi-orthogonal NLFM waveforms with APC. The total ambiguity energies corresponding to cases **a**, **b**, **c**, and **d** are −20.58 dB, −21.39 dB, −24.05 dB, and −24.87 dB, respectively. It should be noted that different color scales are applied across the subplots

the range direction (Fig. 6.7c), achieving about a 3.47 dB reduction in measured ambiguity energy relative to the chirp. The combination of quasi-orthogonal NLFM waveforms with APC further lowers the ambiguity energy by roughly 4.29 dB, with the remaining ambiguities still defocused, as shown in Fig. 6.7d.

To achieve azimuth ambiguity suppression with quasi-orthogonal NLFM waveforms in hybrid and $\pm\pi/4$ quad-pol SAR systems, two modified setups—termed the generalized hybrid and generalized $\pm\pi/4$ quad-pol configurations—are introduced. The timing diagrams for these generalized configurations are shown in Fig. 6.8a and b, respectively.

Figure 6.9 presents the 2-D IRFs for the first azimuth ambiguous region of a generalized hybrid or generalized $\pm\pi/4$ quad-pol SAR (HV channel), assuming equal energy for all raw signals. In Fig. 6.9a, the azimuth ambiguities of the standard chirp waveform appear partially focused. By contrast, the azimuth ambiguities for

Table 6.7 Total measured ambiguity energy for a point target

Ambiguity	Tx scheme	Ambiguity energy[a] (dB)
Range	APC method	−0.81
	Quasi-orthogonal NLFM waveforms	−3.47
	Combination of quasi-orthogonal NLFM waveforms and APC	−4.29
Azimuth	Up chirp waveforms	0
	Up-down chirp waveforms	−0.003
	Quasi-orthogonal NLFM waveforms	−3.48

[a] Total ambiguity energy relative to a chirp signal

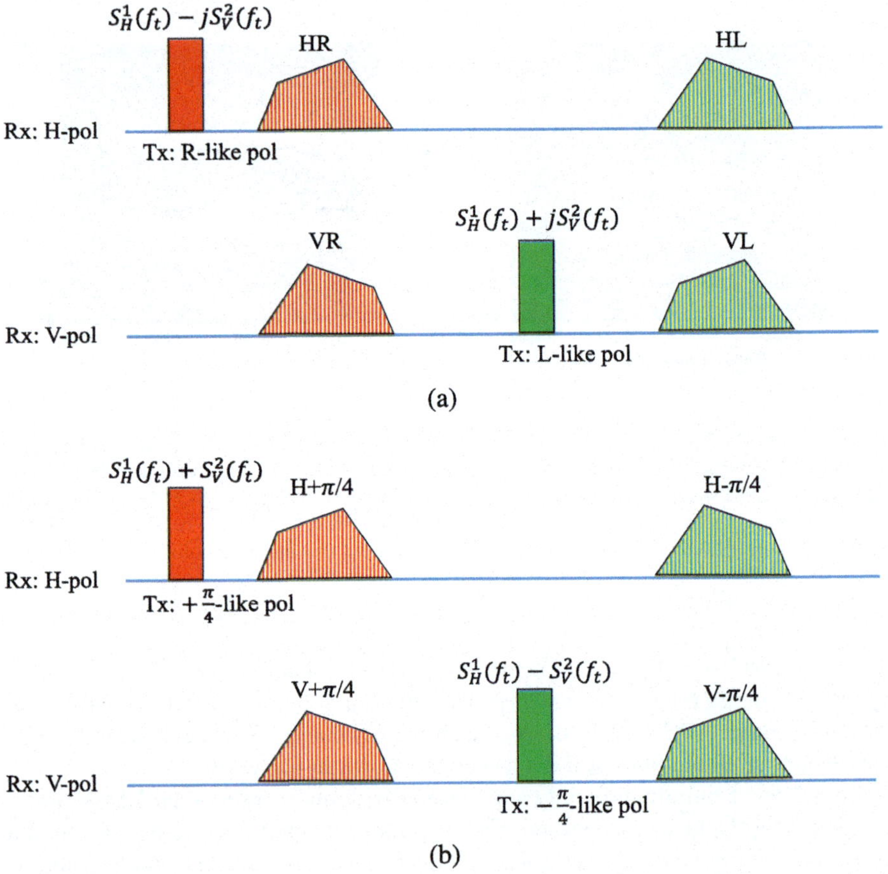

Fig. 6.8 Timing diagrams: **a** Generalized hybrid quad-pol SAR. **b** Generalized $\pm\pi/4$ quad-pol SAR

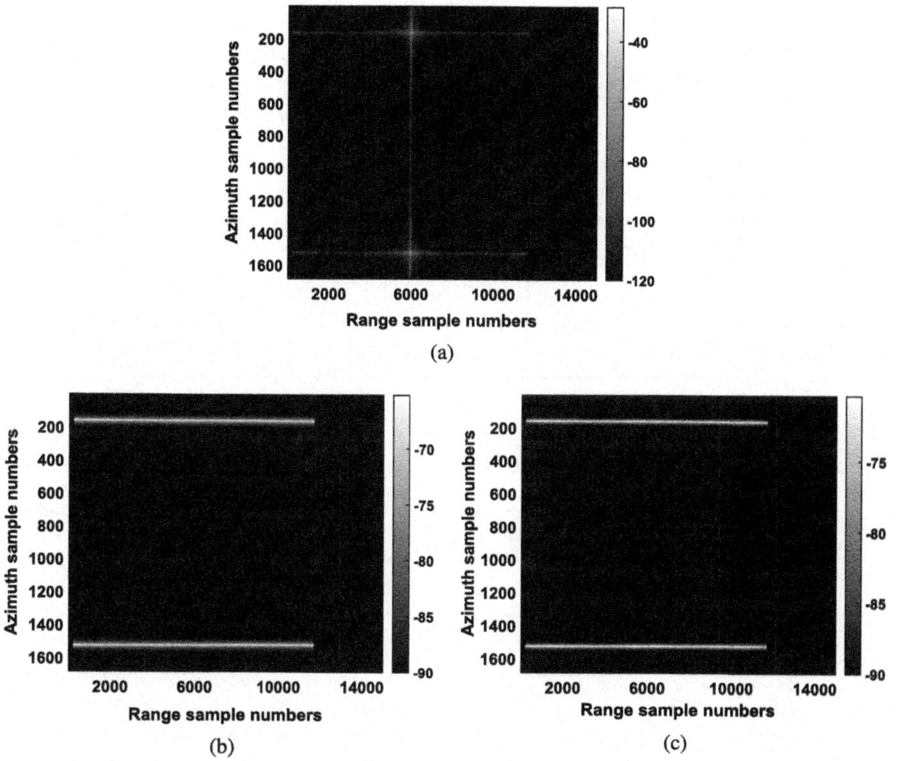

Fig. 6.9 2-D IRFs (in dB) of the first azimuth ambiguous region (HV channel in a generalized hybrid or generalized $\pi/4$ quad-pol mode) are shown for: **a** up chirp waveform, **b** up–down chirp waveforms, and **c** quasi-orthogonal NLFM waveforms. The total ambiguity energies for these cases are −21.98 dB, −21.95 dB, and −25.46 dB, respectively. Note that the color scales differ between subplots

the up–down chirp and quasi-orthogonal NLFM waveforms are defocused, with energy spread along the range direction. Table 6.7 summarizes the measured azimuth ambiguity levels relative to a chirp signal. The total ambiguity energy for the up–down chirp pair is nearly identical to that of the standard chirp, as these waveforms do not reduce the CCE. In comparison, the quasi-orthogonal NLFM waveforms provide an approximate 3.48 dB reduction in azimuth ambiguity energy relative to the chirp.

6.4.3 Distributed Target Simulation

Distributed target simulations are performed using the LT-1 system parameters to study ambiguity interference in the HV polarization. This analysis considers range ambiguities in a conventional quad-pol mode and azimuth ambiguities in generalized

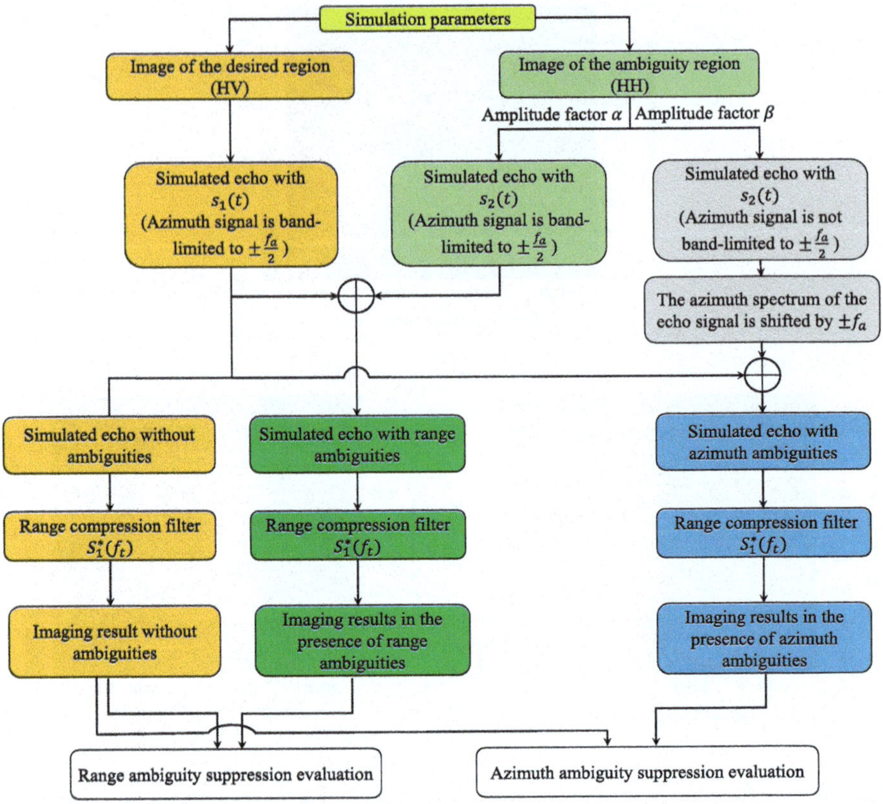

Fig. 6.10 The simulator is set up to assess the suppression of both the first near-range ambiguity (HV channel in a conventional quad-pol system) and the first azimuth ambiguity (HV channel in generalized hybrid or generalized $\pm\pi/4$ quad-pol configurations). Simulations use the LT-1 system parameters, with the azimuth antenna pattern ignored when evaluating range ambiguities

hybrid or generalized $\pm\pi/4$ quad-pol configurations. The simulator for evaluating the first near-range and first azimuth ambiguities is illustrated in Fig. 6.10. Assuming a two-way elevation antenna pattern with a first sidelobe attenuation of −26.6 dB and a backscatter reflectivity difference of 10 dB between HH and HV polarizations ($\sigma_{\text{HH,dB}} - \sigma_{\text{HV,dB}} = 10\,\text{dB}$), the amplitude ratio of the simulated range ambiguity echo is calculated as

$$\alpha = 10^{\frac{-26.6+10}{20}} = 0.1479. \tag{6.36}$$

Figure 6.11 shows simulation results of range ambiguity interference for different transmission and reception schemes, with range along the horizontal axis and azimuth along the vertical axis. Panels (a) and (b) display the desired and ambiguous regions, respectively, while panels (c)–(f) present the imaging outcomes for various Tx/Rx

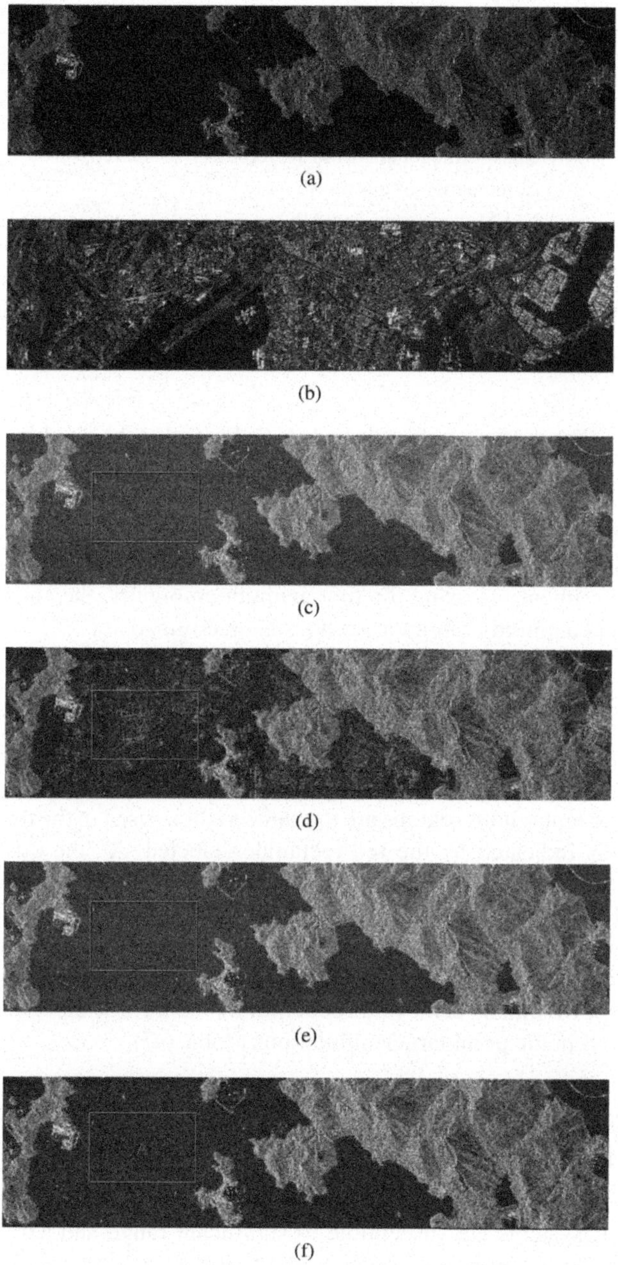

Fig. 6.11 Simulation results for a distributed scene in the HV polarization channel under a conventional quad-pol mode. Panel **a** shows the target area (HV polarization), and panel **b** displays the range-ambiguity region (HH polarization). The imaging outcomes for different approaches are presented in panels (**c**)–(**f**): **c** up–down chirp waveforms, **d** the APC method, **e** quasi-orthogonal NLFM waveforms, and **f** a combined scheme using quasi-orthogonal NLFM waveforms with APC

Table 6.8 Measured total ambiguity energy for distributed target scenarios

Ambiguity	Tx scheme	Ambiguity energy[a] (dB)
Range	Up chirp waveforms	0
	APC method	−0.84
	Quasi-orthogonal NLFM waveforms	−3.51
	Combination of quasi-orthogonal NLFM waveforms and APC	−4.35
Azimuth	Up chirp waveforms	0
	Up-down chirp waveforms	−0.004
	Quasi-orthogonal NLFM waveforms	−3.48

[a] Total ambiguity energy relative to a chirp signal

configurations. The red rectangles marked 'A' indicate that the combination of quasi-orthogonal NLFM waveforms with the APC technique provides the most effective suppression of range ambiguities. Furthermore, Table 6.8 reports the total measured ambiguity energies relative to a standard chirp, which closely match the results obtained from point target simulations (Table 6.7).

For the experiment assessing the first azimuth ambiguity, the amplitude scaling of the simulated azimuth-ambiguous echoes is determined as

$$\beta = 10^{\frac{10}{20}} = 3.16. \tag{6.37}$$

The results corresponding to the different waveforms are displayed in Fig. 6.12, with the horizontal axis representing range and the vertical axis representing azimuth. The desired and ambiguous regions are the same as those used in the range ambiguity simulation.[4] As indicated by the red rectangles labeled 'A', the quasi-orthogonal NLFM pair exhibits superior azimuth ambiguity suppression compared to the other waveforms. The ambiguity energies relative to the chirp signal for Fig. 6.12a, b, and c are summarized in Table 6.8. Additionally, the total measured interference energies for chirp and up-down chirp signals are approximately the same, while the quasi-orthogonal NLFM signals achieve a reduction of −3.48 dB, closely matching the results observed in the point target simulations (Table 6.7).

6.5 Conclusion

Quad-pol SAR systems are susceptible to significant range and azimuth ambiguities due to the relatively lower returned energies in cross-polarized channels compared with co-polarized channels. This study addresses these issues by designing and applying quasi-orthogonal NLFM waveforms, aiming to mitigate range ambiguities

[4] Without loss of generality, both range and azimuth ambiguous returns are generated by appropriately utilizing the backscattering characteristics shown in Fig. 6.11b.

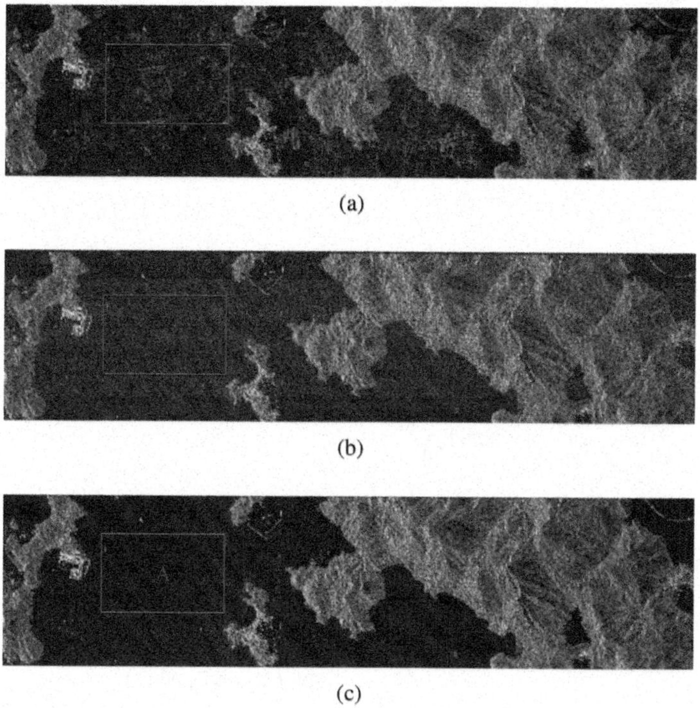

Fig. 6.12 Distributed target simulations are performed for the HV polarization channel in generalized hybrid or generalized $\pi/4$ quad-pol configurations. The resulting images correspond to **a** an up-chirp waveform, **b** up–down chirp waveforms, and **c** quasi-orthogonal NLFM waveforms

in conventional quad-pol SAR and azimuth ambiguities in hybrid/$\pm\pi/4$ quad-pol SAR configurations. The proposed waveforms achieve a reduced CCE compared with existing alternatives, thereby enhancing ambiguity suppression performance. At the same time, they maintain low PSLRs to ensure satisfactory imaging quality. Furthermore, the APC method is integrated to form a combined scheme, which jointly exploits waveform diversity and APC to further suppress range ambiguities in conventional quad-pol mode. Finally, a systematic framework is presented for evaluating AASR and RASR in a conventional quad-pol SAR system employing waveform diversity. To validate the effectiveness of the proposed approach, comprehensive simulations based on the LT-1 system parameters are performed, demonstrating that the designed waveforms and the combined suppression strategies achieve superior performance.

Potential future research may focus on the development and implementation of innovative phase coding techniques for SAR systems. Another promising direction involves designing appropriate calibration methods for hybrid and $\pm\pi/4$ quad-pol SARs when operating with diverse waveforms. This effort can build upon insights gained from recent studies on polarimetric SAR calibration [36–43] and represents a highly relevant research avenue.

6.6 The Derivation of (6.19)

In this part, the derivation of (6.19) is reported.

$$
\begin{aligned}
f_{n,m}^{1,2}(\tau) &= \exp\left(-j\pi\alpha_{1,n}\tau^2\right)\exp\left(j2\pi\left(b_{1,n}-\Delta b_{n,m}^{1,2}\right)\tau\right)\\
&\times \Phi_{n,m}^{1,2}\mathrm{rect}\left(\frac{\tau-(n-m)T_s}{2T_s}\right)\int_{t_1'}^{t_2'}\exp\left(j2\pi\left(\alpha_{1,n}\tau+\Delta b_{n,m}^{1,2}\right)t'\right)dt'\\
&= \exp\left(-j\pi\alpha_{1,n}\tau^2\right)\exp\left(j2\pi\left(b_{1,n}-\Delta b_{n,m}^{1,2}\right)\tau\right)\Phi_{n,m}^{1,2}\\
&\times\left[\mathrm{rect}\left(\frac{\tau-(n-m-1/2)T_s}{T_s}\right)\right.\\
&\times\int_0^{\tau+(1+m-n)T_s}\exp\left(j2\pi\left(\alpha_{1,n}\tau+\Delta b_{n,m}^{1,2}\right)t'\right)dt'\\
&+\mathrm{rect}\left(\frac{\tau-(n-m+1/2)T_s}{T_s}\right)\\
&\times\left.\int_{\tau+(m-n)T_s}^{T_s}\exp\left(j2\pi\left(\alpha_{1,n}\tau+\Delta b_{n,m}^{1,2}\right)t'\right)dt'\right]\\
&= \exp\left(-j\pi\alpha_{1,n}\tau^2\right)\exp\left(j2\pi\left(b_{1,n}-\Delta b_{n,m}^{1,2}\right)\tau\right)\Phi_{n,m}^{1,2}\\
&\times\left[\mathrm{rect}\left(\frac{\tau-(n-m-1/2)T_s}{T_s}\right)\frac{e^{j2\pi\left(\alpha_{1,n}\tau+\Delta b_{n,m}^{1,2}\right)(\tau+(1+m-n)T_s)}-1}{j2\pi\left(\alpha_{1,n}\tau+\Delta b_{n,m}^{1,2}\right)}\right.\\
&+\mathrm{rect}\left(\frac{\tau-(n-m+1/2)T_s}{T_s}\right)\\
&\times\left.\frac{e^{j2\pi\left(\alpha_{1,n}\tau+\Delta b_{n,m}^{1,2}\right)T_s}-e^{j2\pi\left(\alpha_{1,n}\tau+\Delta b_{n,m}^{1,2}\right)(\tau+(m-n)T_s)}}{j2\pi\left(\alpha_{1,n}\tau+\Delta b_{n,m}^{1,2}\right)}\right]\\
&= \exp\left(-j\pi\alpha_{1,n}\tau^2\right)\exp\left(j2\pi\left(b_{1,n}-\Delta b_{n,m}^{1,2}\right)\tau\right)\Phi_{n,m}^{1,2}\\
&\times\exp\left(j\pi\left(\alpha_{1,n}\tau+\Delta b_{n,m}^{1,2}\right)(T_s+(m-n)T_s+\tau)\right)\\
&\times\left[\mathrm{rect}\left(\frac{\tau-(n-m-1/2)T_s}{T_s}\right)(T_s+(m-n)T_s+\tau)\right.\\
&\times\mathrm{sinc}\left(\left(\alpha_{1,n}\tau+\Delta b_{n,m}^{1,2}\right)(T_s+(m-n)T_s+\tau)\right)\\
&+\mathrm{rect}\left(\frac{\tau-(n-m+1/2)T_s}{T_s}\right)(T_s-(m-n)T_s-\tau)\\
&\times\left.\mathrm{sinc}\left(\left(\alpha_{1,n}\tau+\Delta b_{n,m}^{1,2}\right)(T_s-(m-n)T_s-\tau)\right)\right]\\
&= \exp\left(-j\pi\alpha_{1,n}\tau^2\right)\exp\left(j2\pi\left(b_{1,n}-\Delta b_{n,m}^{1,2}\right)\tau\right)\Phi_{n,m}^{1,2}\\
&\times\exp\left(j\pi\left(\alpha_{1,n}\tau+\Delta b_{n,m}^{1,2}\right)(T_s+(m-n)T_s+\tau)\right)\\
&\times(T_s-|\tau+(m-n)T_s|)\,\mathrm{rect}\left(\frac{\tau-(n-m)T_s}{2T_s}\right)\\
&\times\mathrm{sinc}\left(\left(\alpha_{1,n}\tau+\Delta b_{n,m}^{1,2}\right)(T_s-|\tau+(m-n)T_s|)\right).
\end{aligned}
\tag{6.38}
$$

References

1. Levanon N, Mozeson E. Radar signals. Hoboken, NJ, USA: Wiley; 2004.
2. Gini F, De Maio A, Patton L. Waveform design and diversity for advanced radar systems. Stevenage, U.K: Inst. Eng. Technol; 2012.
3. Kerahroodi MA, Aubry A, De Maio A, et al. A Coordinate-Descent Framework to Design Low PSL/ISL Sequences. IEEE Transactions on Signal Processing. 2017;65(22):5942–5956.
4. Wang W. Mitigating Range Ambiguities in High-PRF SAR With OFDM Waveform Diversity. IEEE Geoscience and Remote Sensing Letters. 2013;10(1):101–105.
5. Jin G, Deng Y, Wang R, et al. Mitigating Range Ambiguities With Advanced Nonlinear Frequency Modulation Waveform. IEEE Geoscience and Remote Sensing Letters. 2019;16(8):1230–1234.
6. Krieger G, Gebert N, Moreira A. Multidimensional Waveform Encoding: A New Digital Beamforming Technique for Synthetic Aperture Radar Remote Sensing. IEEE Transactions on Geoscience and Remote Sensing. 2008;46(1):31–46.
7. Knapskog AO. Range Ambiguity Suppression in Space-Borne SAR by Up- and Down-Chirp Modulation in Combination with Pseudo-Random Bi-Phase Coding. In: EUSAR 2018; 12th European Conference on Synthetic Aperture Radar; 2018. p. 1–4.
8. Krieger G. MIMO-SAR: Opportunities and Pitfalls. IEEE Transactions on Geoscience and Remote Sensing. 2014;52(5):2628–2645.
9. Moreira A, Prats-Iraola P, Younis M, et al. A tutorial on synthetic aperture radar. IEEE Geoscience and Remote Sensing Magazine. 2013;1(1):6–43.
10. Cloude S. Polarisation: applications in remote sensing. New York, NY, USA: Oxford university press; 2010.
11. Lee JS, Pottier E. Polarimetric radar imaging: from basics to applications. Boca Raton, FL, USA: CRC press; 2009.
12. Raney RK, Freeman A, Jordan RL. Improved Range Ambiguity Performance in Quad-Pol SAR. IEEE Transactions on Geoscience and Remote Sensing. 2012;50(2):349–356.
13. Raney RK. Hybrid-Polarity SAR Architecture. IEEE Transactions on Geoscience and Remote Sensing. 2007;45(11):3397–3404.
14. Souyris JC, Imbo P, Fjortoft R, et al. Compact polarimetry based on symmetry properties of geophysical media: The $\pi/4$ mode. IEEE Transactions on Geoscience and Remote Sensing. 2005;43(3):634–646.
15. Villano M, Krieger G, Moreira A. New Insights Into Ambiguities in Quad-Pol SAR. IEEE Transactions on Geoscience and Remote Sensing. 2017;55(6):3287–3308.
16. Dall J, Kusk A. Azimuth phase coding for range ambiguity suppression in SAR. In: IGARSS 2004. 2004 IEEE International Geoscience and Remote Sensing Symposium. vol. 3; 2004. p. 1734–1737.
17. Bordoni F, Younis M, Krieger G. Ambiguity Suppression by Azimuth Phase Coding in Multichannel SAR Systems. IEEE Transactions on Geoscience and Remote Sensing. 2012;50(2):617–629.
18. Jin G, Deng Y, Wang W, et al. On the SAR Imaging Performance Analysis of Alternate Transmitting Mode Based on Waveform Diversity: Theory and Simulation. IEEE Geoscience and Remote Sensing Letters. 2020;17(9):1553–1557.
19. Guo L, Tan X, Dang H. Range ambiguity suppression for multi-channel SAR system near singular points. In: 2016 IEEE International Geoscience and Remote Sensing Symposium (IGARSS); 2016. p. 2086–2089.
20. Fowle E. The design of FM pulse compression signals. IEEE Transactions on Information Theory. 1964;10(1):61–67.
21. Vizitiu I, Anton L, Popescu F, et al. The synthesis of some NLFM laws using the stationary phase principle. In: 2012 10th International Symposium on Electronics and Telecommunications; 2012. p. 377–380.
22. Kurdzo JM, Cheong BL, Palmer RD, et al. Optimized NLFM pulse compression waveforms for high-sensitivity radar observations. In: 2014 International Radar Conference; 2014. p. 1–6.

23. Saeedi J, Faez K. Synthetic aperture radar imaging using nonlinear frequency modulation signal. IEEE Transactions on Aerospace and Electronic Systems. 2016;52(1):99–110.
24. Wang W, Wang R, Zhang Z, et al. First Demonstration of Airborne SAR With Nonlinear FM Chirp Waveforms. IEEE Geoscience and Remote Sensing Letters. 2016;13(2):247–251.
25. Wang W, Wang R, Deng Y, et al. Demonstration of NLFM Waveforms With Experiments and Doppler Shift Compensation for SAR Application. IEEE Geoscience and Remote Sensing Letters. 2016;13(12):1999–2003.
26. Jin G, Deng Y, Wang R, et al. An Advanced Nonlinear Frequency Modulation Waveform for Radar Imaging With Low Sidelobe. IEEE Transactions on Geoscience and Remote Sensing. 2019;57(8):6155–6168.
27. Jin G, Liu K, Deng Y, et al. Nonlinear Frequency Modulation Signal Generator in LT-1. IEEE Geoscience and Remote Sensing Letters. 2019;16(10):1570–1574.
28. Wang W. MIMO SAR OFDM Chirp Waveform Diversity Design With Random Matrix Modulation. IEEE Transactions on Geoscience and Remote Sensing. 2015;53(3):1615–1625.
29. Boersma J. Computation of Fresnel integrals. Math Comput. 1960;14(69-72):380.
30. Ugray Z, Lasdon L, Plummer J, et al. Scatter search and local NLP solvers: A multistart framework for global optimization. INFORMS Journal on Computing. 2007;19(3):328–340.
31. Curlander JC, Mcdonough RN. Synthetic Aperture Radar: Systems and Signal Processing. New York, NY, USA: Wiley; 1991.
32. Li FK, Johnson WTK. Ambiguities in Spacebornene Synthetic Aperture Radar Systems. IEEE Transactions on Aerospace and Electronic Systems. 1983;19(3):389–397.
33. Bayma RW. Aperture size and ambiguity constraints for a synthetic aperture radar. In: IEEE 1975 international Radar conference. vol. 21; 1975. p. 499–504.
34. Villano M, Krieger G. Spectral-Based Estimation of the Local Azimuth Ambiguity-to-Signal Ratio in SAR Images. IEEE Transactions on Geoscience and Remote Sensing. 2014;52(5):2304–2313.
35. Ulaby F, Dobson MC, Álvarez-Pérez JL. Handbook of radar scattering statistics for terrain. Dedham, MA, USA: Artech House; 1989.
36. Freeman A, van Zyl JJ, Klein JD, et al. Calibration of Stokes and scattering matrix format polarimetric SAR data. IEEE Transactions on Geoscience and Remote Sensing. 1992;30(3):531–539.
37. Villa A, Iannini L, Giudici D, et al. Calibration of SAR Polarimetric Images by Means of a Covariance Matching Approach. IEEE Transactions on Geoscience and Remote Sensing. 2015;53(2):674–686.
38. Ainsworth TL, Ferro-Famil L, Jong-Sen Lee. Orientation angle preserving a posteriori polarimetric SAR calibration. IEEE Transactions on Geoscience and Remote Sensing. 2006;44(4):994–1003.
39. Sarabandi K. Calibration of a polarimetric synthetic aperture radar using a known distributed target. IEEE Transactions on Geoscience and Remote Sensing. 1994;32(3):575–582.
40. Quegan S. A unified algorithm for phase and cross-talk calibration of polarimetric data-theory and observations. IEEE Transactions on Geoscience and Remote Sensing. 1994;32(1):89–99.
41. Kimura H, Mizuno T, Papathanassiou KP, et al. Improvement of polarimetric SAR calibration based on the Quegan algorithm. In: IGARSS 2004. 2004 IEEE International Geoscience and Remote Sensing Symposium. vol. 1; 2004. p. 187.
42. Shimada M, Isoguchi O, Tadono T, et al. PALSAR Radiometric and Geometric Calibration. IEEE Transactions on Geoscience and Remote Sensing. 2009;47(12):3915–3932.
43. Shimada M. Model-Based Polarimetric SAR Calibration Method Using Forest and Surface-Scattering Targets. IEEE Transactions on Geoscience and Remote Sensing. 2011;49(5):1712–1733.

Chapter 7
MIMO-SAR Waveform Design

7.1 Introduction

Synthetic Aperture Radar (SAR) has become a key remote sensing tool due to its ability to operate in all weather conditions and provide continuous day-and-night observations for large-scale environmental monitoring. With application demands growing and technical specifications becoming more stringent–particularly in terms of achieving higher resolution and broader swath coverage–researchers have been exploring innovative SAR architectures. Among these developments, the Multiple-Input Multiple-Output (MIMO) SAR concept has emerged as a promising solution.

Originally introduced in wireless communications to enhance channel capacity [1], the MIMO concept has demonstrated strong resistance to noise and narrowband fading [2, 3]. When this idea migrated into radar, it gave rise to MIMO radar, which has quickly attracted broad interest because of its considerable potential to boost system capability [4]. In the context of imaging radars, MIMO-SAR has become particularly compelling. By offering more degrees of freedom than conventional SAR architectures, it enables significant improvements in system performance [5–7]. One of its major advantages lies in providing additional spatial samples and effectively extending the synthetic aperture in azimuth, which makes it well suited for applications such as High-Resolution Wide-Swath (HRWS) imaging [8], Ground Moving Target Indication (GMTI) [9, 10], as well as interference and tomography-related tasks [11]. The concept has also inspired novel system architectures, for instance the downward-looking 3-D imaging scheme known as ARTINO, where cross-track resolution is further enhanced [12, 13]. Moreover, compared with traditional multistatic SAR configurations, MIMO-SAR can collect multiple slightly diverse viewing angles over the same area. By jointly processing both monostatic and bistatic observations, it is possible to achieve improved range resolution [14, 15].

In a MIMO-SAR configuration, each receiving antenna collects the ground-scattered signals generated by all transmit channels. For a given transmitter, the returns produced by the remaining transmit antennas act as mutual interference,

R. Wang and W. Wang, *Synthetic Aperture Radar (SAR) Waveform Design and Signal Processing*, https://doi.org/10.1007/978-981-95-7549-7_7

which can significantly impair image quality. Consequently, a key difficulty in practical MIMO-SAR operation lies in effectively separating the echoes associated with different transmitters.

7.1.1 Previous Work

In general, MIMO-SAR echo separation approaches fall into three main categories:

(1) Time-Division Multiplexing (TDM) waveforms [16, 17];
(2) Frequency-Division Multiplexing (FDM) waveforms [18, 19];
(3) Code-Division Multiplexing (CDM) waveforms [20, 21].

However, TDM-based designs broaden the overall pulse duration because of time partitioning, which in turn restricts the achievable swath. FDM-based solutions, on the other hand, demand additional frequency-conversion hardware and tend to suffer from reduced coherence, ultimately raising system complexity and cost. As a result, both TDM and FDM waveform strategies see limited suitability for spaceborne MIMO-SAR systems. In contrast, CDM waveforms are considered a more viable option for such applications.

In early CDM-based waveform designs, echo separation for different transmit channels was achieved by employing mutually orthogonal signals in the time–frequency domain [21, 22]. Krieger [7] later pointed out that such theoretically orthogonal waveforms are unsuitable for practical MIMO-SAR operation, as they conflict with the constraints dictated by Parseval's theorem[1] and because SAR scenes are inherently composed of distributed scatterers.[2] In response to these limitations, research attention moved toward multi-dimensional coding approaches. A representative example is the Space–Time Coding (STC) Alamouti scheme–transmit-side coding combined with receive-side decoding–originally introduced by Kim [23] and subsequently by Wang et al. [21]. Follow-up studies extended this framework to handle time-varying propagation channels [24, 25] and to improve the SNR of the reconstructed echoes [26].

In addition, Orthogonal Frequency-Division Multiplexing (OFDM) chirp signals have been considered a viable option for mitigating interference energy because their discrete subcarriers are inherently orthogonal. Yet this orthogonality holds only when the echo duration matches a specific length. To address this limitation, Kim et al. introduced a Digital Beam Forming (DBF) approach that partitions the full swath into several sub-swaths, thereby shortening the received echo and enabling the orthogonality condition to be fulfilled [27], and an enhanced version of this method was later presented in [28].

On the other hand, OFDM-type waveforms are highly susceptible to Doppler-induced leakage, which compromises their ability to achieve reliable echo separation.

[1] Orthogonal waveforms with flat spectra generate mismatched energy that remains in the image rather than being removed; see Appendix A in [7].

[2] The accumulated mismatched energy from numerous distributed targets becomes non-negligible.

To mitigate this problem, Krieger [7] proposed an orthogonal-waveform beamforming approach that combines orthogonal signal construction with DBF processing. In this scheme, interference from scatterers with small angular separations is suppressed using Short-Term Shift-Orthogonal (STSO) waveforms, while echoes originating from wider angular regions are further attenuated by receive-side DBF. Extending this concept, the Segmental-Shifted Chirp (SSC) beamforming technique presented in [29] provides another viable solution, and its implementation is planned for validation on NASA's airborne DBSAR-2 MIMO-SAR platform.

7.1.2 Main Advantages and Organization

This work proposes a new spaceborne MIMO-SAR imaging framework utilizing Segmented-Phase-Code (SPC) waveforms tailored for M transmit channels, with the objective of alleviating inter-channel echo interference. In this scheme, each transmit waveform is partitioned into several segments, and prescribed phase values are applied within each segment to encode a chosen modulation waveform. The modulation waveform can be flexibly selected, and it determines the resulting range response once echo separation is performed. After the reflected echoes from the scene are collected, signals arriving from nearby angles are distinguished through a simple time-shift weighting operation. For echoes originating from more separated directions, the method employs a combination of range band-pass filtering and Digital Beamforming (DBF). Relative to existing MIMO-SAR strategies, the proposed SPC-based approach offers the following main advantages:

- Compared with the advanced STSO waveform beamforming approach, the proposed method relaxes the strict short-term shift-orthogonality requirement while maintaining imaging quality, thereby broadening the range of usable waveform designs.
- In contrast to other current MIMO-SAR approaches, the proposed method mitigates Doppler leakage, accommodates time-varying channel conditions, and operates without the need to raise the Pulse-Repetition Frequency (PRF).

The structure of the paper is arranged as follows. Section 7.2 first outlines the fundamental issue of echo separation and then introduces the SPC waveform together with the overall spaceborne MIMO-SAR imaging scheme. Section 7.3 examines how the swath extent and the number of band-pass filters affect system performance and further develops a phase-coding design that maximizes the SNR. Section 7.4 reports comprehensive simulation studies that demonstrate the practicality of the proposed method. Comparative evaluations against representative MIMO-SAR imaging techniques are discussed in Sect. 7.5. Section 7.6 closes the paper with concluding remarks.

Notation: Boldface uppercase and lowercase symbols are used for matrices and vectors, respectively. The operators $(\cdot)^T$ and $(\cdot)^H$ denote the transpose and Hermitian transpose. The symbol $*t$ refers to convolution along the fast-time dimension.

Rank(A) stands for the rank of matrix $\boldsymbol{A}$. The element located at the nth row and mth column of $\boldsymbol{A}$ is written as $\boldsymbol{A}_{n,m}$, and $\boldsymbol{I}_k$ denotes the $k \times k$ identity matrix.

7.2 MIMO-SAR Echo Separation Scheme

7.2.1 Simple Overview of the Echo Separation Problem

Figure 7.1 depicts the geometry of a MIMO-SAR configuration, where the Tx/Rx antenna locations are intentionally exaggerated for ease of illustration. Under the go-stop assumption, transmit and receive antennas share the same position; in contrast, a MIMO-SAR system activates all Tx/Rx elements to transmit waveforms and capture the corresponding ground echoes, thereby enlarging the set of available phase centers. For a traditional multi-channel SAR employing M colocated Tx/Rx antennas, at most M phase centers can be formed, whereas a MIMO-SAR setup theoretically yields M^2 pairs. However, many of these phase centers are not unique. As shown in Fig. 7.2, the phase center associated with the path Rx1 $\rightarrow$ Tx2 is identical to that of Tx1 $\rightarrow$ Rx2. As a result, a MIMO-SAR system with M Tx/Rx elements effectively provides only $2M - 1$ distinct phase centers, which supports reduced PRF requirements, enlarged swath coverage, and the formation of extended baselines [7].

However, every receive channel records the superposed echoes originating from all active transmit antennas. Let s_m denote the waveform emitted by the mth transmitter, which can be written as

$$s_m(t) = \text{rect}\left(\frac{t}{T}\right)\exp\left(j\theta_m(t)\right), \tag{7.1}$$

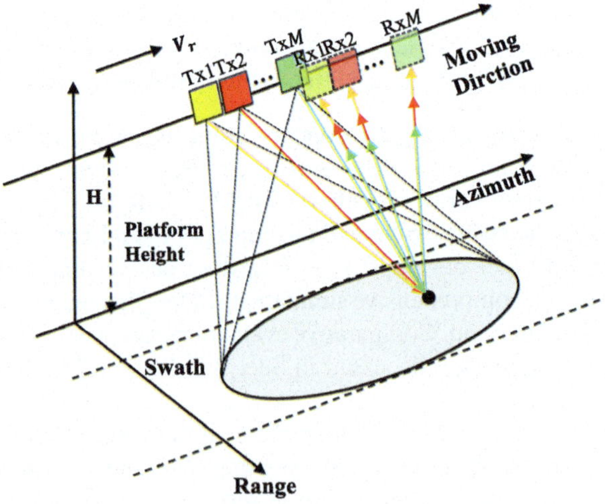

Fig. 7.1 Illustration of the MIMO-SAR system geometry

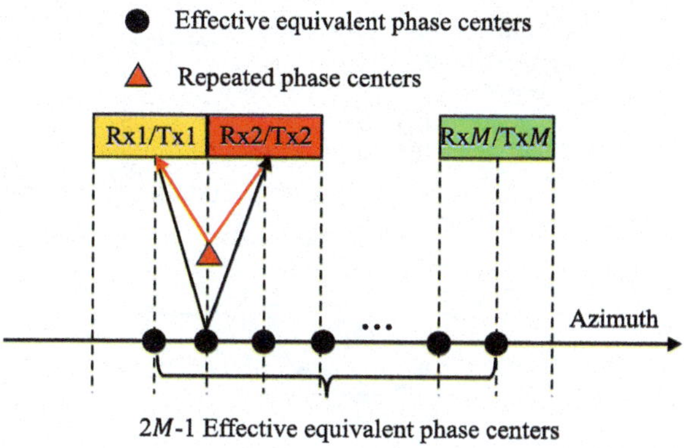

Fig. 7.2 Equivalent phase points offered by a MIMO-SAR system using M Tx/Rx antennas

where t represents fast time, T denotes the pulse duration, and $\theta_m(t)$ corresponds to the phase of the mth transmitted radar signal, and

$$\text{rect}\,(t) = \begin{cases} 1, & -0.5 \le t \le 0.5 \\ 0, & \text{elsewhere} \end{cases} . \tag{7.2}$$

After bouncing off a point target, the returned echo is mixed down to baseband. The component of the received signal corresponding to the waveform emitted by the mth transmitter and captured by the pth receiver can be expressed as

$$\begin{aligned} r_{p,m}(t,\eta) = \sigma_0 \text{rect}\left(\frac{t - t_{p,m}}{T}\right) \exp(-j2\pi f_c t_{p,m}) \\ \times \exp\left(j\theta_m(t - t_{p,m})\right), \end{aligned} \tag{7.3}$$

where σ_0 is the target backscattering coefficient and f_c denotes the carrier frequency. In addition, $t_{p,m}$ corresponds to the two-way propagation delay, which can be further written as

$$t_{p,m} = \frac{R_{p,m}(\eta)}{c}, \tag{7.4}$$

where c denotes the speed of light, and $R_{p,m}(\eta)$ represents the slant-range evolution with respect to the azimuth time η, given by

$$\begin{aligned} R_{p,m}(\eta) &= R_m(\eta) + R_p(\eta) \\ &= \sqrt{R_0^2 + \left(V_r\eta + \left(m - \frac{M+1}{2}\right)\frac{L_a}{M}\right)^2} \\ &+ \sqrt{R_0^2 + \left(V_r\eta + \left(p - \frac{M+1}{2}\right)\frac{L_a}{M}\right)^2}, \end{aligned} \tag{7.5}$$

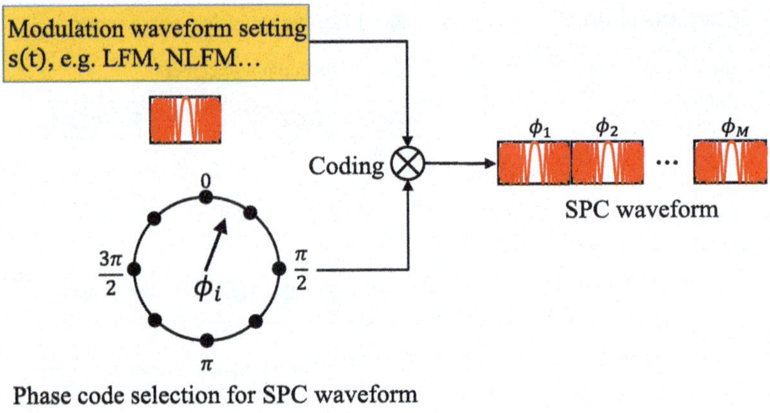

Fig. 7.3 Schematic depiction of SPC waveform generation

where $R_m(\eta)$ denotes the slant-range trajectory between the mth antenna and the target, R_0 is the minimum slant range referenced to the azimuth antenna center $\frac{M+1}{2}$, η represents the slow time relative to the time of closest approach, V_r is the platform velocity, and L_a is the azimuth antenna length. Moreover, (7.3) can be expressed in a more concise form.

$$r_{p,m}(t,\eta) = h_{p,m}(t,\eta) *_t s_m(t), \tag{7.6}$$

where $h_{p,m}(t,\eta)$ denotes the channel response corresponding to the mth transmitter and the pth receiver, which can be written as

$$h_{p,m}(t,\eta) = \sigma_0 \delta\left(t - t_{p,m}(\eta)\right) \exp(-j2\pi f_c t_{p,m}(\eta)) . \tag{7.7}$$

The composite echo captured by the pth antenna can be expressed as

$$r_p(t,\eta) = \sum_{m=1}^{M} r_{p,m}(t,\eta), \tag{7.8}$$

and, for each transmit antenna, the goal of the echo separation process is to obtain its corresponding echo, namely $r_{p,1}(t), r_{p,2}(t), \ldots, r_{p,M}(t)$.

7.2.2 Scheme Description

The orthogonal-waveform beamforming separation scheme primarily involves two steps. First, echoes arriving from closely spaced angles are separated, followed by the application of a spatial filter, specifically the DBF technique, to isolate echoes from more distant angles. To accomplish the first step, carefully designed orthogonal signals, such as OFDM waveforms, are utilized. However, Doppler shifts in the radar can substantially degrade waveform orthogonality, thereby impacting the separation performance, as discussed in [7, 30, 31]. To address this, STSO waveforms have been proposed to mitigate Doppler leakage, although they must satisfy the short-term shift-orthogonality requirement. In contrast, the proposed method does not rely on transmitted waveform orthogonality for echo separation. Instead, it introduces a novel radar waveform, termed the SPC waveform. Its straightforward construction is illustrated in Fig. 7.3, and it can be mathematically expressed as

$$s_{\mathrm{spc}}(t) = \sum_{i=1}^{M} \phi_i s(t-(i-1)T_s), \tag{7.9}$$

where $s(t)$ denotes the sub-pulse used as the modulation waveform, which can be chosen from any radar signal type, such as Linear Frequency Modulation (LFM) or Nonlinear Frequency Modulation (NLFM) [32, 33]. In (7.9), ϕ_i is the constant phase applied to the ith segment of the signal. If $s(t)$ has a pulse duration T_s and a bandwidth B, the resulting SPC waveform will have a total pulse length of $T = MT_s$ while maintaining the same bandwidth B.

Moreover, in a MIMO-SAR system with M transmitters, the SPC waveform allocated to the mth antenna can be written as

$$s_m(t) = \sum_{i=1}^{M} \phi_{m,i} s(t-(i-1)T_s), \tag{7.10}$$

where $\phi_{m,i}$ denotes the phase code applied to the ith segment of the mth transmit antenna. Considering the structure of the transmitted waveforms, (7.6) can be reformulated as

$$r_{p,m} = h_{p,m}(t,\eta) *_t \boldsymbol{\phi_m} \boldsymbol{s}, \tag{7.11}$$

where $\boldsymbol{\phi_m}$ denotes the phase code vector associated with the mth transmitted waveform, defined such that

$$\boldsymbol{\phi_m} = [\phi_{m,1}, \phi_{m,2}, \ldots, \phi_{m,M}], \tag{7.12}$$

and $\boldsymbol{s}$ denotes the vector of the modulation waveform

$$\boldsymbol{s} = [s(t), s(t-T_s), \ldots, s(t-(M-1)T_s)]^T. \tag{7.13}$$

The composite echo $r_p(t,\eta)$ can alternatively be expressed as

$$r_p(t,\eta) = \sum_{m=1}^{M} r_{p,m}(t) = \boldsymbol{h}_p *_t \boldsymbol{\Phi} \boldsymbol{s}, \tag{7.14}$$

where $\boldsymbol{h}_p$ denotes the vector of channel responses

$$\boldsymbol{h}_p = [h_{p,1}, h_{p,2}, \ldots, h_{p,M}], \tag{7.15}$$

and $\boldsymbol{\Phi}$ represents the matrix of phase codes

$$\boldsymbol{\Phi} = \begin{bmatrix} \boldsymbol{\phi}_1 \\ \boldsymbol{\phi}_2 \\ \vdots \\ \boldsymbol{\phi}_M \end{bmatrix}. \tag{7.16}$$

In order to retrieve the echo $r_{p,m}(t)$ at the pth receiver corresponding to the transmitted waveform $s_m(t)$, a straightforward time-shift weighting operation is applied to the combined echo $r_p(t)$, as

$$\begin{aligned} r_{p,m}^{\text{weight}}(t, \eta) &= \sum_{i=1}^{M} w_{m,i} r_p(t - (i-1)T_s, \eta) \\ &= \sum_{i=1}^{M} w_{m,i} \boldsymbol{h}_p *_t \boldsymbol{\Phi} \boldsymbol{s}_i, \end{aligned} \tag{7.17}$$

where $r_p(t - (i-1)T_s, \eta)$ $(i = 1, 2, \ldots, M)$ denotes the ith time-shifted echo, and $\boldsymbol{s}_i$ represents the associated time-shifted modulation waveform vector, as

$$\boldsymbol{s}_i = \begin{bmatrix} s(t - (i-1)T_s) \\ s(t - (i-1)T_s) \\ \vdots \\ s(t - (M-1)T - (i-1)T_s) \end{bmatrix} \tag{7.18}$$

It is evident that the time-shifted echo signals share a common time segment within $(M-1)T_s + t_{p,m} \leq t \leq MT_s + t_{p,m}$. In other words, they contain a shared component, specifically $s(t - (M-1)T_s)$. The time-shifted echo within this interval, $(M-1)T_s + t_{p,m} \leq t \leq MT_s + t_{p,m}$, can be further expressed as

$$
\begin{aligned}
r_{p,m}^{\text{common}}(t,\eta) &= \boldsymbol{h}_p *_t \boldsymbol{\Phi}\left\{\begin{bmatrix} 0 \\ 0 \\ \vdots \\ w_{m,1}s(t-(M-1)T_s) \end{bmatrix}\right. \\
&+ \begin{bmatrix} 0 \\ \vdots \\ w_{m,2}s(t-(M-1)T_s) \\ 0 \end{bmatrix} + \cdots \\
&+ \left.\begin{bmatrix} w_{m,M}s(t-(M-1)T_s) \\ \vdots \\ 0 \\ 0 \end{bmatrix}\right\} \\
&= \boldsymbol{h}_p *_t \{s(t-(M-1)T_s)\boldsymbol{\Phi}\boldsymbol{w}_m\},
\end{aligned}
\tag{7.19}
$$

where $\boldsymbol{w}_m = [w_{m,M}, w_{m,M-1}, \ldots, w_{m,1}]^T$ represents the vector of weighting coefficients used to separate the echo signal $r_{p,m}(t,\eta)$. In addition, since neither $\boldsymbol{\Phi}$ nor $\boldsymbol{w}_m$ depends on the variable t, (7.19) can be reformulated as

$$
r_{p,m}^{\text{common}}(t,\eta) = s(t-(M-1)T_s) *_t \boldsymbol{h}_p\boldsymbol{\Phi}\boldsymbol{w}_m. \tag{7.20}
$$

It is clear that $\boldsymbol{w}_m$ must meet the following condition

$$
\boldsymbol{w}_m = \boldsymbol{\Phi}^{-1}\boldsymbol{c}_m, \tag{7.21}
$$

where $\boldsymbol{c}_m$ represents the desired weighted outcome within $(M-1)T_s + t_{p,m} \le t \le M \cdot T_s + t_{p,m}$ for isolating the echo $r_{p,m}(t,\eta)$, as

$$
\boldsymbol{c}_m = [\underbrace{0,\ldots,0}_{m-1}, 1, \ldots, 0]^T. \tag{7.22}
$$

As shown in (7.21), an effective weighting coefficient vector $\boldsymbol{w}_m$ can be obtained only if the phase code matrix $\boldsymbol{\Phi}$ is invertible. That is, within the interval $(M-1)T_s + t_{p,m} \le t \le MT_s + t_{p,m}$, the application of a properly designed weighting vector $\boldsymbol{w}_m$ enables retention of the target echo $r_{p,m}(t)$ while simultaneously suppressing interfering signals, specifically

$$
\begin{aligned}
&r_{p,m}^{\text{weight}}(t,\eta) = r_{p,m}(t,\eta), \\
&(M-1)T_s + t_{p,m} \le t \le MT_s + t_{p,m}
\end{aligned}
\tag{7.23}
$$

For a system with $M = 2$ transmit antennas, Fig. 7.4 provides a simplified illustration of the echo separation process. After applying the time-shift weighting procedure, the separated echo $r_{p,m}$ can be recovered within the overlapping time interval.

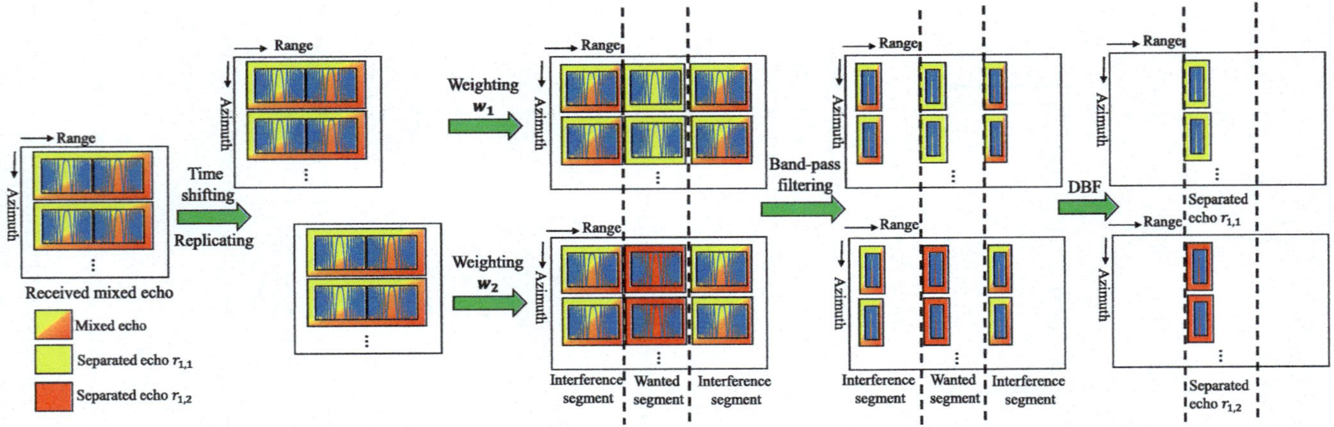

Fig. 7.4 Schematic illustration of the echo separation procedure for the 1st receive antenna ($M = 2$), with range migration effects ignored. Here, only a single band-pass filtered output from one receive channel is displayed; the full-bandwidth result can be reconstructed by aggregating all filtered outputs

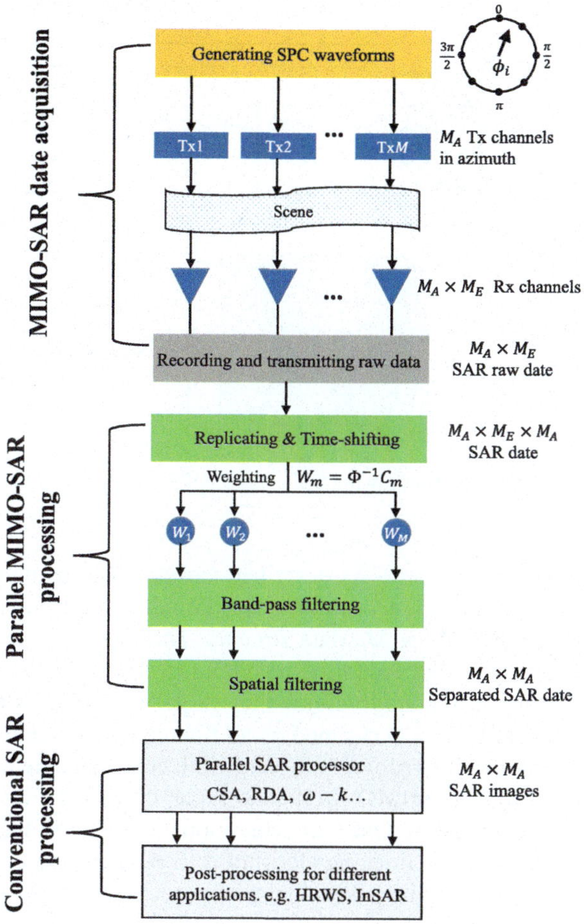

Fig. 7.5 Overview of a MIMO-SAR approach based on SPC waveforms

Digital Beamforming (DBF) is then employed to attenuate interference from echoes arriving at more distant angles. Nevertheless, because the desired signal segment partially overlaps in time with interference segments, DBF alone is insufficient for full separation. To mitigate this, additional band-pass filters are introduced to reduce Pulse Extension Loss (PEL). By applying band-pass filtering, both the target and interference echoes are divided into multiple narrowband components in the time domain, and interference suppression is further improved using DBF. The role of the band-pass filter is conceptually similar to a matched filter as discussed in [7], yet it is considerably easier to implement and can significantly reduce the cost of a spaceborne MIMO-SAR system.

Furthermore, Fig. 7.5 depicts the general framework of a MIMO-SAR system employing SPC waveforms, while Fig. 7.6 shows the antenna configuration. Let M_{A} and M_{E} represent the number of antennas in the azimuth and elevation direc-

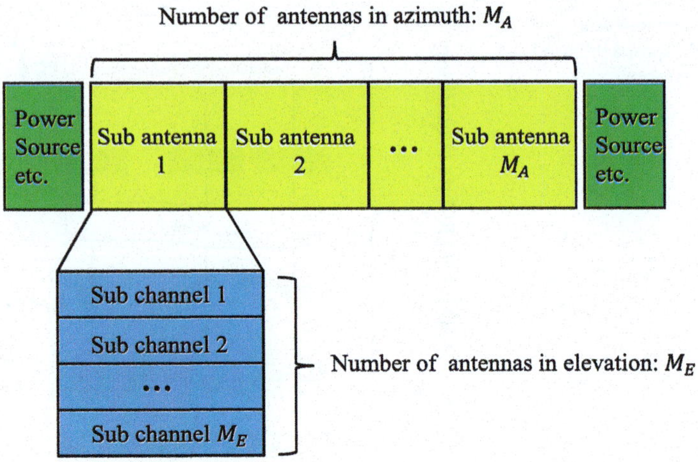

Fig. 7.6 Diagram showing the antenna unit

tions, respectively. As described earlier, each azimuth antenna transmits a unique SPC waveform, and all antennas simultaneously receive ground-reflected echoes, producing $M_\mathrm{A} \times M_\mathrm{E}$ data streams. These echoes are then replicated, subjected to time-shifts, and weighted, resulting in $M_\mathrm{A} \times M_\mathrm{E} \times M_\mathrm{A}$ streams. Next, band-pass filtering is applied to divide the echoes into multiple narrowband segments. Subsequently, DBF is performed on all filtered outputs, and the narrowband segments are recombined to recover full-bandwidth echo data, yielding $M_\mathrm{A} \times M_\mathrm{A}$ separated echo streams. The MIMO-SAR echo separation can be carried out in parallel across all receive channels. Once the separation is completed, the data are processed using conventional SAR imaging algorithms to generate $M_\mathrm{A} \times M_\mathrm{A}$ SAR images. Depending on the application, additional post-processing procedures may be applied as needed.

7.2.3 Spatial Filter Design

This section provides a detailed description of the spatial filter design. After applying time-shift weighting, the echo is divided into several time segments using N_B band-pass filters. Here, an up-chirp waveform is assumed as the modulation signal of the SPC waveforms without loss of generality, and the result from the n_Bth band-pass filter can be written as

$$
\begin{aligned}
r_{\mathrm{BF}}(t, \eta) &= \operatorname{rect}\left(\frac{t - t_{p,m} - T_{\mathrm{want}}(n_B)}{\frac{T_s}{N_B}}\right) r_{p,m}^{\mathrm{weight}}(t, \eta) \\
&+ \sum_{i=-M+1, i\neq 0}^{M-1} \operatorname{rect}\left(\frac{t - t_{p,m} - T_{\mathrm{want}}(n_B) - iT_s}{\frac{T_s}{N_B}}\right) r_{p,m}^{\mathrm{weight}}(t, \eta) \\
&= r_{\mathrm{want}}(t, \eta) + r_{\mathrm{inf}}(t, \eta),
\end{aligned} \tag{7.24}
$$

where, $T_{\mathrm{want}}(n_B)$ indicates the time shift with respect to the center of the desired signal pulse, and can be written as

$$
T_{\mathrm{want}}(n_B) = \frac{(M-2)T_s}{2} + \frac{(2n_B - 1)T_s}{2N_B}. \tag{7.25}
$$

Besides,

$$
r_{\mathrm{want}}(t, \eta) = \operatorname{rect}\left(\frac{t - t_{p,m} - T_{\mathrm{want}}(n_B)}{\frac{T_s}{N_B}}\right) r_{p,m}^{\mathrm{weight}}(t, \eta), \tag{7.26}
$$

and according to (7.23), $r_{p,m}^{\mathrm{weight}}(t) = r_{p,m}(t)$ in $(M-1)T_s + t_{p,m} \leq t \leq MT_s + t_{p,m}$. Therefore, (7.26) can be rewritten as

$$
r_{\mathrm{want}}(t, \eta) = \operatorname{rect}\left(\frac{t - t_{p,m} - T_{\mathrm{want}}(n_B)}{\frac{T_s}{N_B}}\right) r_{p,m}(t, \eta), \tag{7.27}
$$

which corresponds to the extracted portion of the desired signal, and

$$
\begin{aligned}
r_{\mathrm{inf}}(t, \eta) = &\sum_{i=-M+1, i\neq 0}^{M-1} \\
&\operatorname{rect}\left(\frac{t - t_{p,m} - T_{\mathrm{want}}(n_B) - i \times T_s}{\frac{T_s}{N_B}}\right) r_{p,m}^{\mathrm{weight}}(t, \eta),
\end{aligned} \tag{7.28}
$$

represents the interference echo that the spatial filter is intended to suppress. The actual arrival angle of each echo is determined by the target's slant range, i.e., the signal time delay. The delay between the desired echo and the ith interfering echo is iT_s, implying that they originate from different elevation angles. In a DBF system with N sub-antennas along the elevation direction, the steering vectors associated with the desired and interfering echoes can be formulated as

$$
\boldsymbol{v}_0(k) = [e^{jd_1 k_0}, \ldots, e^{jd_N k_0}]^T, \tag{7.29}
$$

respectively, and

$$\begin{aligned} \boldsymbol{v}_i(k) &= [e^{jd_1 k_i}, \ldots, e^{jd_N k_i}]^T, \\ i &= [-M+1, \ldots, M-1], i \neq 0 \end{aligned} \tag{7.30}$$

where $\lambda = c/f_c$ is the wavelength, N denotes the number of DBF channels, $d_n = \left(n - \frac{N+1}{2}\right) d$ where d is the spacing between elevation channels, and k_0 is the wave number of the desired echo.

$$k_0 = \frac{2\pi}{\lambda} \sin(\varphi_0), \tag{7.31}$$

while k_i represents the wave number of the ith interfering echo.

$$k_i = \frac{2\pi}{\lambda} \sin(\varphi_i), \tag{7.32}$$

where φ_0 and φ_i represent the off-boresight angles associated with the time delay terms in (7.26) and (7.28), respectively. In this study, DBF is realized using the Linear Constraint Minimum Variance (LCMV) approach[3] [34]. Based on this, the steering vector matrix can be constructed as

$$\boldsymbol{V} = [\boldsymbol{v}_0, \boldsymbol{v}_{-M+1}, \ldots, \boldsymbol{v}_{M-1}, \frac{d\boldsymbol{v}_{-M+1}}{dk_{-M+1}}, \ldots, \frac{d\boldsymbol{v}_{M-1}}{dk_{M-1}}], \tag{7.33}$$

where

$$\begin{aligned} \frac{d\boldsymbol{v}_i}{dk_i} &= [jd_1 e^{jd_1 k_i}, \ldots, jd_N e^{jd_N k_i}]^T, \\ i &= [-M+1, \ldots, M-1], i \neq 0 \end{aligned} \tag{7.34}$$

The weighting coefficients can be determined as

$$\boldsymbol{w}_{\text{DBF}} = \boldsymbol{g}(\boldsymbol{V}^H \boldsymbol{V})^{-1} \boldsymbol{V}^H, \tag{7.35}$$

where $\boldsymbol{g} = [1, 0, \ldots, 0]$ (containing $4M - 4$ zeros) represents the constraint vector.

7.3 Performance Analysis

This section investigates how the swath width and the number of band-pass filters influence system performance. Additionally, a phase coding strategy optimized for maximizing the SNR is introduced.

---

[3] To enhance interference suppression, first-order derivative constraints are imposed on the interference steering vectors.

7.3.1 Swath Width

First, for a scene with a swath width smaller than the spatial length of a sub-pulse (i.e., $\frac{T_s c}{2}$), the resulting scene echo after range matched filtering and time-shift weighting can be represented as

$$r_{\text{scene}}(t) = \int_{t_{min}}^{t_{min}+W_s} \sum_{i=1}^{M} w_{m,i} Q_p(t - t_{p,m} - (i-1)T_s) dt_{p,m}, \tag{7.36}$$

where $Q_p(t)$ denotes the impulse response of $r_p(t)$, consisting of a series of point-target responses; its precise mathematical expression is not required for the present analysis. Here, $t_{\min}$ ($t_{\min} + W_s$) corresponds to the delay of the nearest (farthest) range cell. Moreover, assuming that $W_s < T_s$, (7.36) can be expressed in the following form

$$r_{\text{scene}}(t) = \int_{t_{min}}^{t_{min}+W_s} Q_{p,m}(t - t_{p,m} - (i-1)T_s) dt_{p,m}, \quad (M-1)T_s + t_{min} \le t \le MT_s + t_{min} \tag{7.37}$$

where $Q_{p,m}(t)$ denotes the impulse response of $r_{p,m}(t)$. This implies that for a short-swath scene, the mixed echoes can be separated effectively without the need for additional DBF. A simplified example for $M = 2$ is shown in Fig. 7.7, where only a single type of signal segment occurs within the shared time interval. The echoes within this interval are then weighted to achieve separation. Similarly, DBF is unnecessary for other previously proposed MIMO-SAR approaches [7, 27, 30, 31], because the OFDM waveform orthogonality is preserved when the swath width is smaller than $\frac{T_s c}{2}$.

When the swath width exceeds $\frac{T_s c}{2}$, multiple types of signal segments appear within the common time interval. In this case, aliasing arises between different sub-pulse echoes, breaking the demodulation condition and ultimately preventing successful separation. An example for $M = 2$ is illustrated in Fig. 7.8. As highlighted by the yellow block, the end portion of the first sub-pulse echo overlaps with the initial part of the second, resulting in aliasing that hampers the separation process. Since spaceborne SAR systems typically operate with swath widths larger than the sub-pulse duration, DBF becomes an indispensable component in spaceborne MIMO-SAR configurations.

7.3.2 SNR Performance

This sub-section incorporates the influence of noise and further examines how to design an optimal phase-coding matrix.

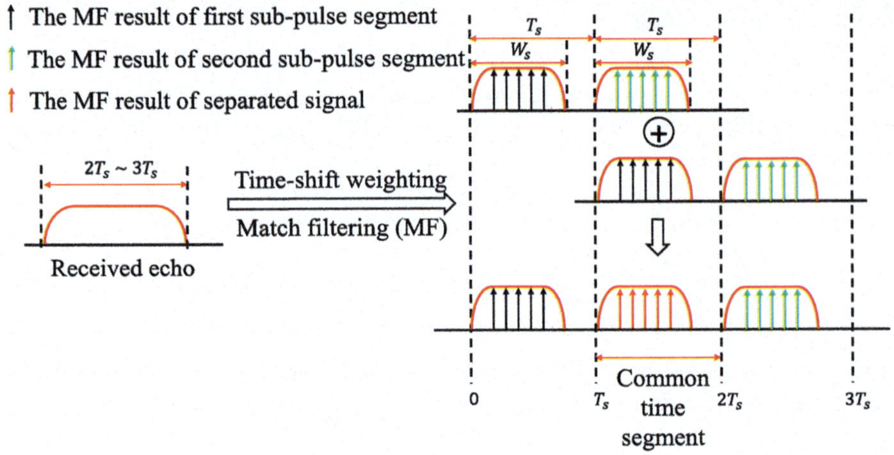

Fig. 7.7 Schematic of echo separation for $M = 2$ when the swath width is less than $\frac{T_s c}{2}$

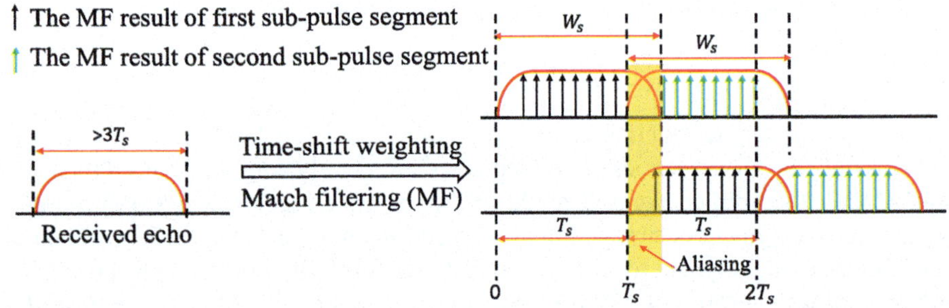

Fig. 7.8 Illustration of the echo-separation procedure for the case of $M = 2$ when the swath width exceeds $\frac{T_s \cdot c}{2}$

The mixed echo at the receiver, denoted by $\tilde{r}_p(t, \eta)$, together with the additive noise term $n_p(t, \eta)$, can be written as

$$\tilde{r}_p(t, \eta) = r_p(t, \eta) + n_p(t, \eta). \tag{7.38}$$

According to the discussion in Sect. 7.2.1 (see (7.23)) and after applying the separation operations to $\tilde{r}_p(t, \eta)$, the echo in the common time segment after time shifting and weighting can be expressed as

$$\hat{r}_{p,m}^{\text{weight}}(t, \eta) = r_{p,m}(t, \eta) + \boldsymbol{n}'_p(t, \eta)\boldsymbol{w}_m, \tag{7.39}$$

where

$$\boldsymbol{n}'_p(t, \eta) = \left[n'_{p,1}(t, \eta), n'_{p,2}(t, \eta), \ldots, n'_{p,M}(t, \eta)\right], \tag{7.40}$$

and $n'_{p,i}(t,\eta)$ $(i=\{1,2,\ldots,M\})$ refers to the noise component corresponding to the ith time-shifted echo within the common time interval.

$$\begin{aligned} &n'_{p,i}(t,\eta)=n_p(t-(i-1)T_s,\eta),\\ &(M-1)T_s+t_{p,m}\le t\le M\cdot T_s+t_{p,m} \end{aligned} \tag{7.41}$$

It should be emphasized that the terms $n'_{p,i}(t,\eta)$ are identically and independently distributed, as each corresponds to a distinct time segment of $n_p(t,\eta)$. To enhance the SNR, the optimal weighting vector can be obtained by casting the problem as the following constrained optimization task:

$$\begin{cases} \min\limits_{\boldsymbol{w}_m} \sum_{i=1}^{M} |w_{m,i}|^2 \\ \text{s.t.}\quad \boldsymbol{\phi}_m \boldsymbol{w}_m = 1, \quad \boldsymbol{w}_m \in \mathbb{C}^M \end{cases} \tag{7.42}$$

where the constraint $\boldsymbol{\phi}_m \boldsymbol{w}_m = 1$ follows directly from (7.21). With this constraint in place, the following inequality (i.e., the Cauchy–Schwarz inequality) can be applied:

$$1=\left|\sum_{i=1}^{M}\phi_{m,i}w_{m,i}\right|^2\le\sum_{i=1}^{M}|\phi_{m,i}|^2\sum_{i=1}^{M}|w_{m,i}|^2=M\sum_{i=1}^{M}|w_{m,i}|^2. \tag{7.43}$$

In other words, the optimal weighting vector is obtained when it is set as

$$\boldsymbol{w}_m^{\text{opt}}=\frac{1}{M}\boldsymbol{\phi}_m{}^H, \tag{7.44}$$

This configuration achieves the minimum of the objective function, specifically

$$\sum_{i=1}^{M}\left|w_{m,i}^{\text{opt}}\right|^2=\frac{1}{M}, \tag{7.45}$$

In this case, the noise energy is decreased to $1/M$, and the energy of the separated echo[4] is similarly reduced to $1/M$, indicating that the weighting procedure does not alter the SNR.

Moreover, the following equation can be readily derived from (7.21):

$$\boldsymbol{W}_m^{\text{opt}}=\left[\boldsymbol{w}_1^{\text{opt}},\boldsymbol{w}_2^{\text{opt}},\ldots,\boldsymbol{w}_M^{\text{opt}}\right]=\boldsymbol{\Phi}^{-1}. \tag{7.46}$$

Additionally, based on (7.44), (7.46) can be rewritten as

$$\frac{1}{M}\boldsymbol{\Phi}^H=\boldsymbol{\Phi}^{-1}. \tag{7.47}$$

[4] Note that the pulse duration of the preserved signal is also reduced to $1/M$.

Therefore, in order to prevent any SNR reduction, the matrix $\frac{1}{\sqrt{M}}\boldsymbol{\Phi}$ needs to be unitary, such as a Fourier matrix.

7.3.3 *The Number of Band-Pass Filters*

As previously discussed, because signal echoes are continuous, band-pass filtering must be applied before DBF to ensure that the desired and interference signal segments correspond to specific time intervals. The width of each signal segment after filtering decreases as the number of band-pass filters increases. In the extreme case, with a sufficiently large number of filters, each segment approaches a single point, effectively acting like a matched filter. At the same time, the DBF suppression performance tends to decrease as the area of suppression expands, so increasing the number of band-pass filters generally enhances interference suppression. However, a higher number of filters also raises system cost and complexity. Therefore, selecting an appropriate number of band-pass filters requires balancing system performance and cost, and a more detailed analysis is provided in the following section.

7.4 Simulation Experiment

To verify the feasibility of the proposed spaceborne MIMO-SAR method, detailed simulation experiments are carried out in this section. A typical spaceborne MIMO-SAR system with two azimuth antenna apertures is considered, and the key system parameters are summarized in Table 7.1. Both azimuth channels transmit distinct SPC waveforms simultaneously, while all antennas collect the ground-reflected echoes concurrently. Additionally, Fig. 7.9 shows an example of the transmitted SPC waveform spectra together with the corresponding phase code matrix.

$$\boldsymbol{\Phi} = \begin{bmatrix} e^{j\frac{3}{2}\pi} & e^{j\frac{1}{3}\pi} \\ & \\ e^{j\frac{2}{3}\pi} & e^{j\frac{1}{2}\pi} \end{bmatrix}. \tag{7.48}$$

It should be noted that the transmitted SPC waveforms are not mutually orthogonal, as emphasized in the enlarged portion of the figure.

Figure 7.10 shows the received signals at the outputs of the first and sixth band-pass filters. The yellow and black segments correspond to interference echoes from the first and second SPC waveforms, respectively, whereas the red segment represents the desired echo retained after time-shift weighting. The horizontal blue dotted lines indicate the illuminated swath. In both cases, following band-pass filtering, the desired and interference echoes are separated by $T_s = 20\,\mu\text{s}$. In this simulation, the DBF approach described in Sect. 7.2.3 is applied after filtering to suppress inter-

Table 7.1 Spaceborne MIMO-SAR system parameters

Parameters	Value
Orbit height	607 Km
Look angle	20 deg
Center frequency	9.6 GHz
Pulse duration	40 μs
Bandwidth	30 MHz
PRF	1723 Hz
Antenna length in azimuth	9.6 m
Antenna length in elevation	3 m
Number of channels in azimuth	2
Number of channels in elevation	16
Number of band-pass filters	6
Illumination swath	16–24 deg

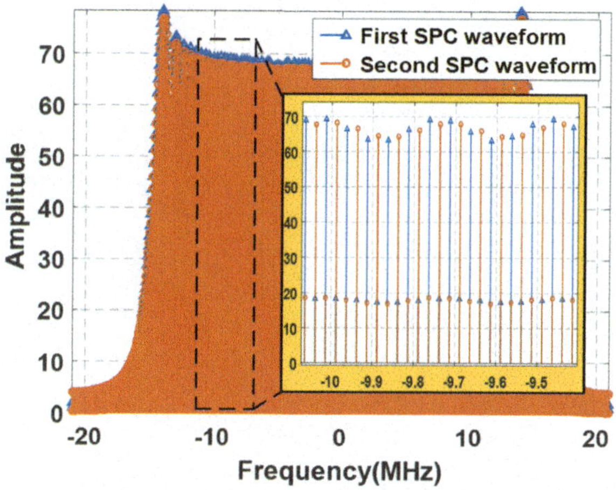

Fig. 7.9 Spectral representation of the SPC waveforms

ference from other antennas. An example receive pattern is shown in Fig. 7.11. For band-pass filter No. 4 (highlighted in red), the peak gain occurs at a Time of Arrival (ToA) of 4.3 ms and an Angle of Arrival (AoA) of 19.57°. According to (7.28), the interference segments have ToAs of 4.3 ms ± 20 μs with AoAs of 18.8° and 20.3°, which align with the nulls of the DBF pattern (yellow and black segments in Fig. 7.11), thereby achieving effective interference suppression. It is also worth noting that the waveform type of the separated signal segment (red) depends on which waveform is preserved during the time-shift weighting process.

Fig. 7.10 Angles of arrival (AoAs) of the received signals after the first (**a**) and sixth (**b**) band-pass filters. The blue dotted lines indicate the illuminated swath. The red segments correspond to the desired echoes, whereas the yellow and black segments represent interference echoes

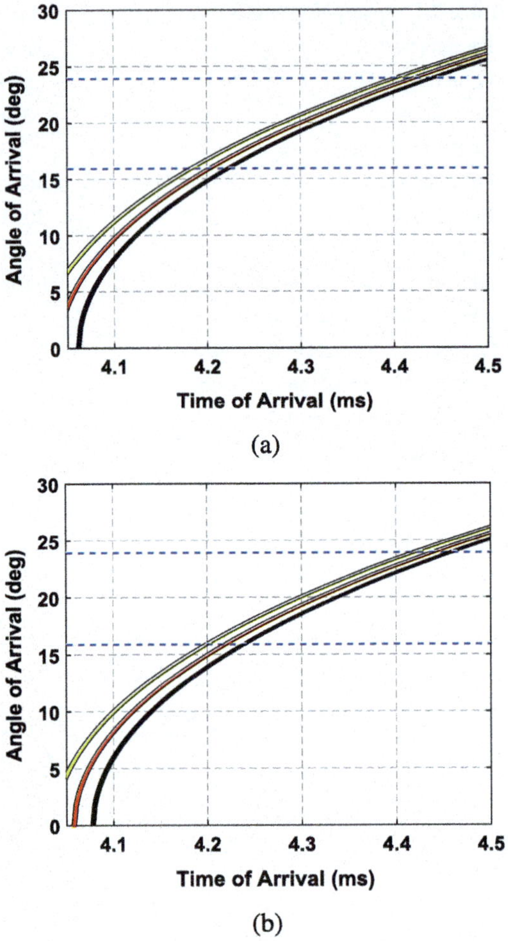

Fig. 7.11 DBF antenna pattern for band-pass filter No. 4 at a ToA of 4.3 ms. The red segment indicates the preserved desired signal, while the interference segments, shown in yellow and black, are effectively attenuated

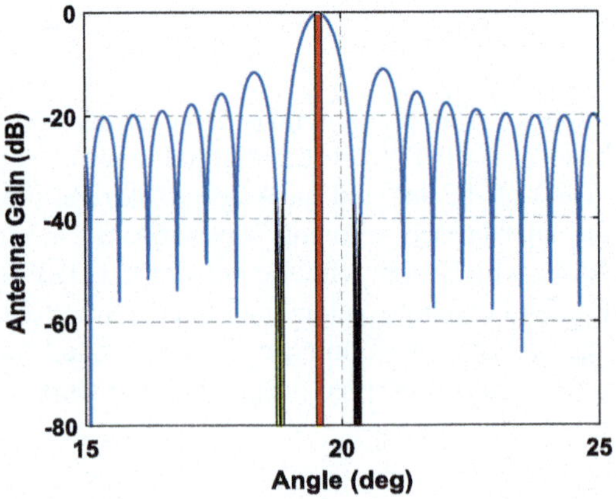

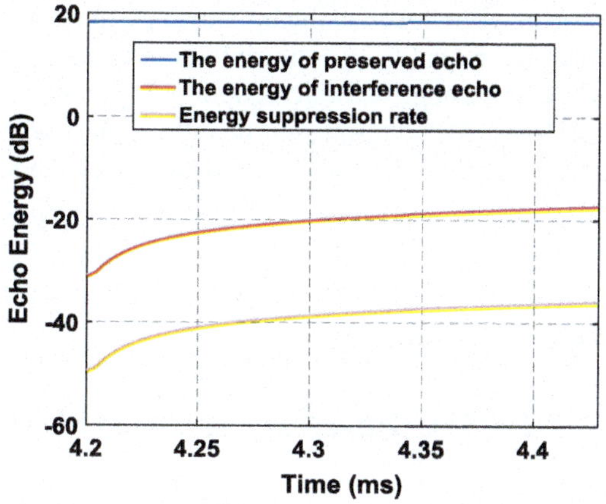

Fig. 7.12 Energy levels of the preserved and interference echoes following DBF processing

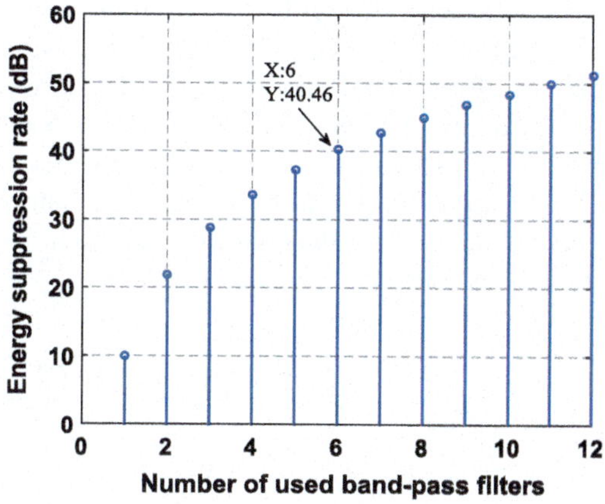

Fig. 7.13 Suppression of interference energy for band-pass filter No. 4 at 4.3 ms as a function of the number of band-pass filters

Figure 7.12 shows the energy levels after DBF for band-pass filter No. 3 at each time instant. Notably, the energy of the interference echoes is suppressed by over 35 dB compared with the preserved signal, which is roughly equivalent to the antenna's cross-polarization coupling. This degree of suppression is adequate for MIMO-SAR operations.

To further evaluate the effect of the number of band-pass filters, simulations were performed to assess interference suppression under different filter counts, with the results shown in Fig. 7.13. The results indicate that suppression performance improves as the number of filters increases, supporting the previous analysis. In particular, employing six band-pass filters yields approximately 40.46 dB of interference suppression, which meets practical system requirements.

To further verify the advantages and effectiveness of the proposed spaceborne MIMO-SAR imaging scheme, a distributed scene simulation was conducted using the same parameters listed in Table 7.1. Figure 7.14a and b show the original scene images used as inputs for the echo simulation of the 1st and 2nd SPC waveforms, respectively.

The imaging result using up–down chirp waveforms is presented in Fig. 7.14c, where the image quality is noticeably poor because the interference energy spreads across the entire image without reduction. In contrast, Fig. 7.14d illustrates that the proposed MIMO-SAR solution significantly suppresses the interference energy, producing an image quality close to that of the original scene. Additionally, the imaging result for the 2nd SPC waveform, shown in Fig. 7.14e, is also well focused.

7.5 Comparisons

7.5.1 SPC Waveform Versus OFDM Waveform

In this section, a comparison between SPC waveforms and OFDM waveforms is presented. Note that the OFDM waveforms considered here are not the traditional ones; instead, they correspond to an improved version that eliminates additional carrier frequency offsets. The conventional OFDM waveform described in [27] represents the standard form, while the improved OFDM waveform in [30] addresses only the compensation for two waveforms. A more general method for generating the improved OFDM waveform has been introduced in our previous work [35]. The mathematical expressions for both the conventional and improved OFDM waveforms are provided directly as follows.

$$s_m^{\text{Conventional}}(t) = e^{j2\pi \frac{m-1}{MT_s}t} \sum_{i=1}^{M} s(t-(i-1)T_s) \tag{7.49}$$

and

$$s_m^{\text{Improved}}(t) = \sum_{i=1}^{M} e^{j\frac{2\pi}{M}(i-1)(m-1)} s(t-(i-1)T_s), \tag{7.50}$$

respectively.

As shown in (7.49), conventional OFDM waveforms incorporate a small additional carrier frequency offset $\frac{m-1}{MT_s}$, which is avoided in (7.50). It can also be observed that the structure of the improved OFDM waveform is closely analogous to that of the SPC waveform, and the method proposed in this paper is inspired by this improved OFDM design. Therefore, it is necessary to examine the relationship between SPC and OFDM waveforms. Moreover, the phase code matrix associated with the improved OFDM waveform is

$$\mathbf{\Phi}_{m,n} = e^{j\frac{2\pi}{M}(n-1)(m-1)}. \tag{7.51}$$

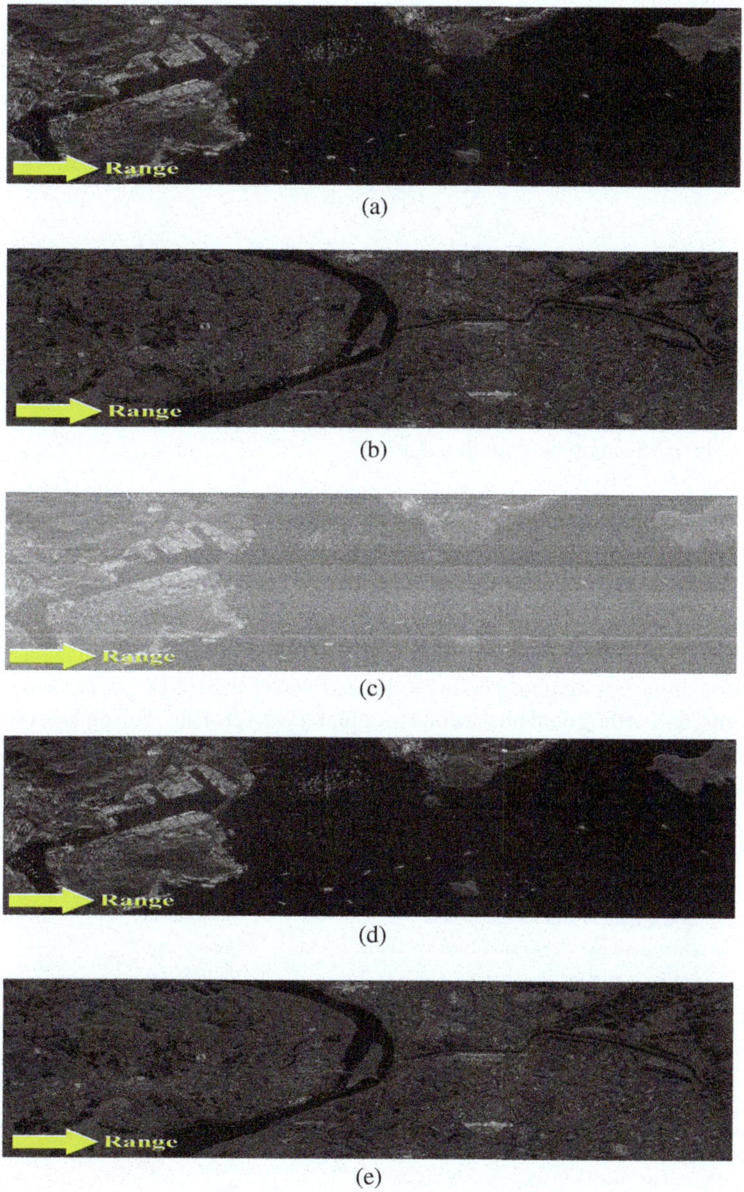

Fig. 7.14 Simulation results for a distributed scene using a MIMO-SAR system with two transmit antennas. The horizontal axis corresponds to the range direction, and the vertical axis corresponds to the azimuth direction. **a** Original scene illuminated by the 1st SPC waveform. **b** Original scene illuminated by the 2nd SPC waveform. **c** Imaging results obtained with up–down chirp waveforms. **d** Reconstructed image of the 1st SPC waveform after echo separation using the proposed method. **e** Reconstructed image of the 2nd SPC waveform after echo separation using the proposed method

Letting $\boldsymbol{E} = \frac{1}{M}\boldsymbol{\Phi}\boldsymbol{\Phi}^H$, $\boldsymbol{E}_{m,n}$ can be calculated as

$$\begin{aligned}\boldsymbol{E}_{m,n} &= \frac{1}{M}\sum_{k=1}^{M} e^{j\frac{2\pi}{M}(n-1)(m-1)} e^{-j\frac{2\pi}{M}(n-1)(m-1)} \\ &= \frac{1}{M}\sum_{k=1}^{M} e^{j\frac{2\pi}{M}(n-m)(k-1)},\end{aligned} \tag{7.52}$$

If $n \neq m$, (7.52) can be calculated as

$$\boldsymbol{E}_{m,n} = \frac{1}{M}\frac{1-1}{1-e^{j\frac{2\pi}{M}(n-m)}} = 0, \tag{7.53}$$

If $n = m$, (7.52) can be calculated as

$$\boldsymbol{E}_{m,n} = \frac{1}{M}\boldsymbol{E}_{n,n} = \frac{1}{M}\sum_{k=1}^{M} 1 = 1 \tag{7.54}$$

Therefore, $\boldsymbol{E}$ is an identity matrix, which indicates that the OFDM waveform can be viewed as a specific case of the SPC waveform. Additionally, Fig. 7.15 depicts the relationships among phase code modulated waveforms, SPC waveforms, OFDM waveforms, and orthogonal phase code modulated waveforms. The phase code matrix

Waveform	Phase code matrix $\frac{1}{\sqrt{M}}\boldsymbol{\Phi}$
Phase code modulated waveform	Arbitrary matrix
SPC waveform for the echo separation	Invertible matrix
SPC waveform for a maximum SNR	Unitary matrix
Orthogonal phase code modulated waveform	Unitary matrix
OFDM waveform	$\Phi_{m,n} = e^{j\frac{2\pi}{M}(n-1)(m-1)}$

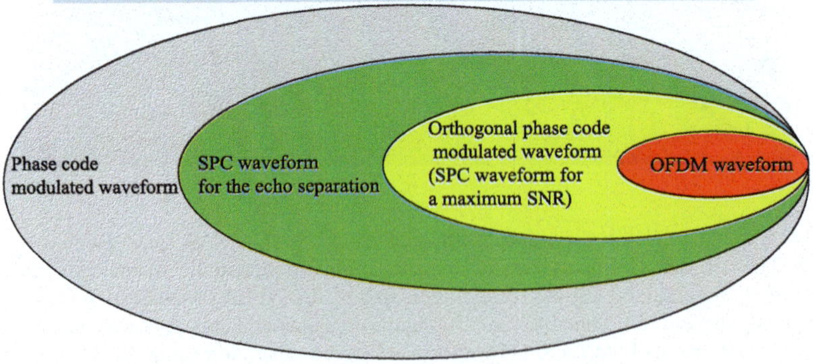

Fig. 7.15 Relationship among phase-coded waveforms, SPC waveforms, OFDM waveforms, and orthogonal phase-coded waveforms

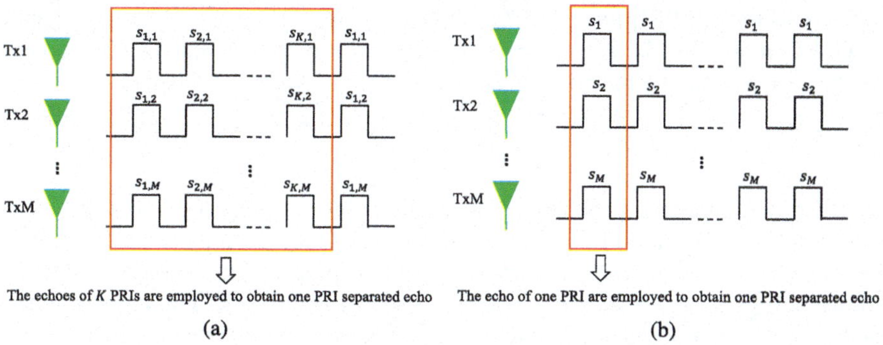

Fig. 7.16 Comparison of transmission schemes: **a** STC MIMO-SAR, **b** Proposed MIMO-SAR method

of OFDM waveforms must satisfy (7.51), classifying them as a type of orthogonal phase code modulated waveform. Ignoring SNR constraints, the sole requirement for the SPC waveform's phase code matrix is invertibility, implying that orthogonal phase code modulated waveforms are naturally a subset of SPC waveforms. Moreover, to achieve maximum SNR, SPC waveforms effectively become orthogonal phase code modulated waveforms.

7.5.2 Comparisons Between Different MIMO-SAR Schemes

This section presents a comprehensive comparison between the proposed approach and both the STC schemes and the orthogonal-waveform beamforming methods.

Figure 7.16a illustrates the transmission strategy of the STC MIMO-SAR system, and its corresponding signal model can be expressed as in [26].

$$
\begin{aligned}
\boldsymbol{R_n}(f_t, f_\eta) &= \begin{bmatrix} R_{n,1}(f_t, f_\eta) \\ R_{n,2}(f_t, f_\eta) \\ \vdots \\ R_{n,K}(f_t, f_\eta) \end{bmatrix} \\
&= \boldsymbol{S}(f_t, f_\eta)\boldsymbol{H_n}(f_t, f_\eta) + \boldsymbol{N_n}(f_t, f_\eta) \qquad (7.55) \\
&= \begin{bmatrix} S_{1,1} & S_{1,2} & \cdots & S_{1,M} \\ S_{2,1} & S_{2,2} & \cdots & S_{2,M} \\ \vdots & \vdots & \ddots & \vdots \\ S_{K,1} & S_{K,2} & \cdots & S_{K,M} \end{bmatrix} \begin{bmatrix} H_{1,n} \\ H_{2,n} \\ \vdots \\ H_{M,n} \end{bmatrix} + \begin{bmatrix} N_{1,n} \\ N_{2,n} \\ \vdots \\ N_{M,n} \end{bmatrix},
\end{aligned}
$$

where $R_{n,i}$ represents the echo received by the nth antenna during the ith Pulse Repetition Interval (PRI), $S_{i,j}(f_t, f_\eta)$ denotes the waveform transmitted by the jth antenna at the ith PRI, and $H_{i,n}(f_t, f_\eta)$ ($N_{i,n}(f_t, f_\eta)$) corresponds to the channel response (noise) for the ith transmit antenna and nth receive antenna. To separate the echoes, a decoding matrix $\boldsymbol{D}$ is applied, which is required to satisfy

$$\boldsymbol{D} \cdot \boldsymbol{S} = \boldsymbol{I}_M \tag{7.56}$$

The requirements specified by (7.56) are Rank($\boldsymbol{D}$) = Rank($\boldsymbol{S}$) = M and $K \geq M$ [36], indicating that the PRF must be greater than $M B_a$ (B_a represents the Doppler bandwidth). For illustration, the transmission scheme of the proposed method is shown in Fig. 7.16b. Unlike the conventional approach, this method does not rely on multiple PRI measurements to obtain a separated echo within a single PRI.

The core concept of orthogonal-waveform beamforming schemes is based on a limited short-term shift-orthogonality requirement [7], specifically,

$$\int h_{\text{DBF}}(\tau) \cdot s_i(t) \cdot s_j^*(t-\tau)dt = 0, \forall \tau \in \mathbb{R}, i \neq j, \tag{7.57}$$

where $s_i(t)$ represents the waveform transmitted by the ith antenna, and $h_{\text{DBF}}(\tau) = \text{rect}(\tau/\Delta\tau)$ denotes a spatial filter. The DBF serves to suppress interferences from signals arriving at distant angles, while conventional schemes require the transmitted waveforms to be orthogonal within $\Delta\tau$ to separate echoes from nearby angles. In contrast, the proposed approach replaces this orthogonality condition with a straightforward time-shift weighting procedure.

7.5.3 Further Discussion Based on Functional Analysis Theory

In this subsection, to clearly highlight the key ideas and contributions of this work, insights from functional analysis theory are applied based on the discussion in Sect. 7.5.2. The MIMO-SAR system can be modeled as an Abstract Algebra System (AAS), with the echo separation processing interpreted as an operator acting on this AAS

$$\Gamma_{\boldsymbol{\Omega}}: \begin{cases} \Gamma_{\boldsymbol{\Omega}}\left\{\bar{s}_i(\boldsymbol{\omega}; \eta; \tau; f_D), h_j(\boldsymbol{\Omega})\right\} = \boldsymbol{0}, \\ \qquad \forall i \neq j, \tau \in \mathbb{T}_\tau, f_D \in \mathbb{F}_D; \\ \\ \Gamma_{\boldsymbol{\Omega}}\left\{\bar{s}_i(\boldsymbol{\omega}; \eta; \tau; f_D), h_j(\boldsymbol{\Omega})\right\} = \chi_i(t; \eta; \tau; f_D), \\ \qquad \forall i = j, \tau \in \mathbb{T}_\tau, f_D \in \mathbb{F}_D; \end{cases} \tag{7.58}$$

where $i, j = 1, \ldots, M$ (with M representing the total number of echoes), τ (f_D) indicates the echo delay (Doppler frequency) relative to the time of closest approach, and $\mathbb{T}_\tau$ and $\mathbb{F}_\eta$ denote the associated domains, which can be expressed as

$$\mathbb{T}_\tau = [\tau^{\min}, \tau^{\max}], \mathbb{T}_\tau \subseteq \mathbb{R}$$
$$\mathbb{F}_D = [f_\eta^{\min}, f_\eta^{\max}], \mathbb{F}_D \subseteq \mathbb{R} \tag{7.59}$$

where $\tau^{\min}$ and $f_\eta^{\min}$ indicate the minimum echo delay and Doppler frequency, and $\tau^{\max}$ and $f_\eta^{\max}$ denote their respective maximum values. The term $\bar{s}_i(\omega; \eta; \tau; f_D)$ refers to the echo corresponding to the ith transmitted signal $s_i(\boldsymbol{\omega})$, $h_j(\boldsymbol{\Omega})$ is the generalized matched function for the jth echo, and $\chi_i(t; \eta; \tau; f_D)$ represents the generalized matched output of the ith echo. Here, $\boldsymbol{\omega}$ specifies the domain of the transmitted waveform, while $\boldsymbol{\Omega}$ defines the processing domain of the operator $\Gamma_{\boldsymbol{\Omega}}$, with $\boldsymbol{\omega} \subseteq \boldsymbol{\Omega}$. In SAR systems, radar signals are transmitted to extract the delay and Doppler frequency embedded in the received echoes; that is, the operator $\Gamma_{\boldsymbol{\Omega}}$ is applied in $\boldsymbol{\Omega}$ to recover the target information τ and f_D. The spaces corresponding to transmitted waveforms and matched functions are denoted by $\mathbb{S}$ and $\mathbb{H}$, respectively.

$$s_i \in \mathbb{S}, h_i \in \mathbb{H}; \mathbb{S}, \mathbb{H} \subseteq \boldsymbol{L}^2, \tag{7.60}$$

respectively.

For a conventional single-input single-output SAR system, the transmitted waveforms are functions of the fast time t, with the operator $\Gamma_{\boldsymbol{\Omega}}$ representing a 2-D time correlation. In this case, $\boldsymbol{\omega} = t$ and $\boldsymbol{\Omega} = (t, \eta)$. Moreover, since SAR is an imaging radar, the waveform space $\mathbb{S}$ includes signals appropriate for SAR imaging, such as LFM and NLFM waveforms, denoted specifically as $\mathbb{S}_{\text{SAR}}$.

Next, for a MIMO-SAR system using orthogonal waveform schemes, the transmitted signals continue to depend on the fast time t, with the operator $\Gamma_{\boldsymbol{\Omega}}$ still corresponding to a 2-D time correlation. However, the transmitted waveform space is

$$\left\{ s_i | s_i \in \mathbb{S}_{\text{SAR}}, \int s_i(t) s_j^*(t - \tau) = 0, \forall \tau \in \mathbb{R}, i \neq j \right\}. \tag{7.61}$$

However, when the waveforms have overlapping spectral content, orthogonality cannot be maintained for arbitrary time shifts.

For STC schemes, the transmitted waveforms are functions of both fast time t and slow time η. MIMO-SAR echo separation can be performed in the Doppler frequency domain, as described in [26]. In this case, $\boldsymbol{\omega} = (t, \eta)$ and $\boldsymbol{\Omega} = (t, \eta, f_D)$, with $\Gamma_{\boldsymbol{\Omega}}$ carrying out the diagonalization of the echo matrix (see (7.56)). The transmitted waveform space is expressed as $\mathbb{S}_{\text{SAR}} \times \mathbb{M}_\eta$, where $\mathbb{M}_\eta$ denotes the azimuth coding space (e.g., Alamouti code), and "×" indicates the Cartesian product. Since a portion of the Doppler information is utilized for echo separation, some Doppler content is lost. For example, in a MIMO-SAR system with two transmit antennas, roughly half of the Doppler information is consumed, leading to a decrease in PRF.

For STSO orthogonal beamforming schemes, DBF introduces an additional observation variable, the direction of arrival φ. Accordingly, $\boldsymbol{\omega} = t$, $\boldsymbol{\Omega} = (t, \eta, \varphi)$, and $\Gamma_{\boldsymbol{\Omega}}$ corresponds to a weighted correlation. While STC schemes exploit Doppler information for MIMO-SAR separation, this same information is also part of the echo of interest, so its use can impair radar observation performance. In contrast, orthogonal-waveform beamforming schemes rely on φ information, which is consequently lost in the final output; fortunately, φ is not the target information. Furthermore, the precision of φ is limited by the number of elevation channels. To overcome this, STSO waveforms are designed to ensure orthogonality among echoes that DBF cannot distinguish in φ, and the transmitted waveform space is

$$\left\{ s_i | s_i \in \mathbb{S}_{\mathrm{SAR}}, \int h_{\mathrm{DBF}}(\tau) \cdot s_i(t) \cdot s_j^*(t-\tau)dt = 0, \right. \\ \left. \forall \tau \in \mathbb{R}, i \neq j \right\}, \tag{7.62}$$

In the proposed approach, a new variable–the phase code ϕ, is introduced through the phase code matrix. It is important to note that ϕ does not serve as an additional observation variable. Its role is to relax the short-term shift-orthogonality limitation. Accordingly, $\boldsymbol{\omega} = (t, \phi)$, $\boldsymbol{\Omega} = (t, \phi, \eta, \varphi)$, and the transmitted waveform space is expressed as $\mathbb{S}_{\mathrm{SAR}} \times \mathbb{C}\phi$ with the optimal phase code space.[5]

$$\mathbb{C}_\phi = \{ \phi | \phi \in \mathbb{C}, \\ \left. \text{The phase code matrix } \frac{1}{\sqrt{M}} \boldsymbol{\Phi} \text{ is a unitary matrix} \right\} \tag{7.63}$$

In this context, compared with the orthogonal-waveform beamforming schemes, the available waveform space is expanded to $\mathbb{S}_{\mathrm{SAR}}$, and the corresponding operator $\Gamma_{\boldsymbol{\Omega}}$ performs time-shift weighted beamforming.

Radar waveform design is a critical factor influencing radar system performance and has been the focus of extensive research, resulting in numerous advanced waveforms aimed at enhancing radar capabilities. Typically, these waveforms do not inherently satisfy the short-term orthogonality requirement. For simpler waveforms, such as LFM signals, STSO versions can be constructed with minimal processing while preserving performance. However, for complex high-performance imaging waveforms (e.g., [32, 33, 37]), enforcing short-term orthogonality adds design constraints, increases complexity, and may degrade waveform quality. The method presented in this paper relaxes this limitation, enabling the direct use of all advanced SAR-suitable waveforms in MIMO-SAR systems, thereby improving performance metrics such as SNR and sidelobe suppression.

Conversely, as noted in Sect. 7.2.3, the receive-side DBF must reliably separate echo signals with a time delay difference of T_s. The DBF conditions are equivalent to those required in the STSO waveform beamforming approach, indicating that the height of an optimally aligned antenna, h_{ant}, needs to satisfy [7].

$$h_{\mathrm{ant}} \geq \frac{2\lambda R_{\max} \tan\left(\theta_{\mathrm{inc,\,max}}\right)}{cT_s}, \tag{7.64}$$

[5] To achieve maximum SNR, $\frac{1}{\sqrt{M}}\boldsymbol{\Phi}$ should be unitary; otherwise, invertibility is sufficient.

Table 7.2 Evaluation of various MIMO-SAR approaches

Method	Main idea of echo separation	Advantages	Drawbacks
STC scheme proposed in [21]	Separated in the slow time domain by a coding and decoding matrix	1. No requirement for DBF	1. PRF $> MB_a$ 2. Residual ambiguous caused by the time-variant channel response
STC scheme proposed in [24]	Separated in the range-Doppler domain by a coding and decoding matrix	1. No requirement for DBF 2. Avoid the effect of the time-variant channel response	1. PRF $> MB_a$ 2. SNR loss
STC scheme proposed in [26]	Separated in the range-Doppler domain by a coding and decoding matrix	1. No requirement for DBF 2. Avoid the effect of the time-variant channel response 3. Avoid the SNR loss in [24]	1. PRF $> MB_a$
OFDM waveform beamforming schemes proposed in [27, 30]	OFDM waveforms, and DBF on receive	1. No requirement for the increase of PRF 2. Avoid the effect of the time-variant channel response	1. Need DBF 2. Suffer from the effect of Doppler leakage
STSO waveform beamforming schemes proposed in [7, 29]	STSO waveforms, and DBF on receive	1. No requirement for the increase of PRF 2. Avoid the effect of the time-variant channel response 3. Avoid the effect of Doppler leakage	1. Need DBF 2. Need to meet short-term shift-orthogonality condition
The method proposed in this paper	Time-shift weighting process and DBF on receive	1. No requirement for the increase of PRF 2. No requirement for the orthogonality of transmitted waveform 3. Avoid the effect of the time-variant channel response 4. Avoid the effect of Doppler leakage	1. Need DBF

where $R_{\max}$ denotes the maximum range distance and $\theta_{\text{inc, max}}$ represents the largest incident angle. A comparison of various MIMO-SAR schemes is presented in Table 7.2.

7.6 Summary

In this work, the SPC waveform, which is a type of phase-coded signal, is employed as the transmitted waveform to address the echo separation problem in MIMO-SAR systems. Compared with conventional STSO waveform beamforming methods, this approach relaxes the short-term shift-orthogonality requirement, thereby expanding the available waveform space. The function of STSO waveforms in separating echoes from closely spaced arrival angles is replaced by a simple time-shift weighting procedure. Additionally, echoes from more distant directions are separated using a combination of range band-pass filtering and DBF. Compared with existing MIMO-SAR methods, the proposed scheme reduces Doppler leakage and the effects of time-varying channels, without increasing the PRF. Detailed numerical simulations are conducted to verify the feasibility of the method, and the relationships between different waveform types as well as performance comparisons among MIMO-SAR schemes are presented.

References

1. A. J. Paulraj and T. Kailath, “Increasing capacity in wireless broadcast systems using distributed transmission/directional reception (dtdr),” Sept. 6 1994. US Patent 5,345,599.
2. I. Khan, M. H. Zafar, M. T. Jan, J. Lloret, M. Basheri, and D. Singh, “Spectral and Energy Efficient Low-Overhead Uplink and Downlink Channel Estimation for 5G Massive MIMO Systems,” *Entropy*, vol. 20, no. 2, 2018.
3. I. Khan, “Notice of Violation of IEEE Publication Principles: A Robust Signal Detection Scheme for 5G Massive Multiuser MIMO Systems,” *IEEE Transactions on Vehicular Technology*, vol. 67, pp. 9597–9604, 2018.
4. E. Fishler, A. Haimovich, R. Blum, D. Chizhik, L. Cimini, and R. Valenzuela, “MIMO radar: an idea whose time has come,” in *Proceedings of the 2004 IEEE Radar Conference (IEEE Cat. No.04CH37509)*, pp. 71–78, April 2004.
5. M. Younis, G. Krieger, and A. Moreira, “MIMO SAR techniques and trades,” in *2013 European Radar Conference*, pp. 141–144, Oct 2013.
6. W. Wang, “MIMO SAR imaging: Potential and challenges,” *IEEE Aerospace and Electronic Systems Magazine*, vol. 28, pp. 18–23, 2013.
7. G. Krieger, “MIMO-SAR: Opportunities and Pitfalls,” *IEEE Transactions on Geoscience and Remote Sensing*, vol. 52, pp. 2628–2645, 2014.
8. D. Cerutti-Maori, I. Sikaneta, J. Klare, and C. H. Gierull, “MIMO SAR Processing for Multichannel High-Resolution Wide-Swath Radars,” *IEEE Transactions on Geoscience and Remote Sensing*, vol. 52, pp. 5034–5055, 2014.
9. J. H. G. Ender, “Along-track array processing for MIMO-SAR/MTI,” in *7th European Conference on Synthetic Aperture Radar*, pp. 1–3, June 2008.
10. W. Zhou, H. Kang, J. Wang, and X. Li, “Effect analysis and velocity estimation of moving target based on MIMO SAR system,” in *2012 IEEE 11th International Conference on Signal Processing*, vol. 3, pp. 1751–1755, Oct 2012.
11. G. Krieger, T. Rommel, and A. Moreira, “MIMO-SAR Tomography,” in *Proceedings of EUSAR 2016: 11th European Conference on Synthetic Aperture Radar*, pp. 1–6, June 2016.
12. J. Klare, M. Weiss, O. Peters, A. Brenner, and J. Ender, “ARTINO: A New High Resolution 3D Imaging Radar System on an Autonomous Airborne Platform,” in *2006 IEEE International Symposium on Geoscience and Remote Sensing*, pp. 3842–3845, July 2006.

13. J. Klare, D. Cerutti-Maori, A. Brenner, and J. Ender, "Image quality analysis of the vibrating sparse MIMO antenna array of the airborne 3D imaging radar ARTINO," in *2007 IEEE International Geoscience and Remote Sensing Symposium*, pp. 5310–5314, July 2007.
14. C. Prati and F. Rocca, "Improving slant-range resolution with multiple SAR surveys," *IEEE Transactions on Aerospace and Electronic Systems*, vol. 29, pp. 135–143, 1993.
15. D. Cristallini, D. Pastina, and P. Lombardo, "Exploiting MIMO SAR Potentialities With Efficient Cross-Track Constellation Configurations for Improved Range Resolution," *IEEE Transactions on Geoscience and Remote Sensing*, vol. 49, pp. 38–52, 2011.
16. R. Boute, "On The Equivalence of Time-Division and Frequency-Division Multiplexing," *IEEE Transactions on Communications*, vol. 33, pp. 97–99, 1985.
17. L. Lo Monte, B. Himed, T. Corigliano, and C. J. Baker, "Performance analysis of time division and code division waveforms in co-located MIMO," in *2015 IEEE Radar Conference (RadarCon)*, pp. 0794–0798, May 2015.
18. J. Xu, G. Liao, S. Zhu, L. Huang, and H. C. So, "Joint Range and Angle Estimation Using MIMO Radar With Frequency Diverse Array," *IEEE Transactions on Signal Processing*, vol. 63, pp. 3396–3410, 2015.
19. F. Bordoni, G. Krieger, and M. Younis, "Multifrequency Subpulse SAR: Exploiting Chirp Bandwidth for an Increased Coverage," *IEEE Geoscience and Remote Sensing Letters*, vol. 16, pp. 40–44, 2019.
20. Hai Deng, "Polyphase code design for Orthogonal Netted Radar systems," *IEEE Transactions on Signal Processing*, vol. 52, pp. 3126–3135, 2004.
21. W. Wang, "Space–Time Coding MIMO-OFDM SAR for High-Resolution Imaging," *IEEE Transactions on Geoscience and Remote Sensing*, vol. 49, pp. 3094–3104, 2011.
22. J. Mittermayer and J. M. Martinez, "Analysis of range ambiguity suppression in SAR by up and down chirp modulation for point and distributed targets," in *IGARSS 2003. 2003 IEEE International Geoscience and Remote Sensing Symposium. Proceedings (IEEE Cat. No.03CH37477)*, vol. 6, pp. 4077–4079 vol.6, July 2003.
23. J. Kim and W. Wiesbeck, "Investigation of a New Multifunctional High Performance SAR System Concept Exploiting MIMO Technology," in *IGARSS 2008 - 2008 IEEE International Geoscience and Remote Sensing Symposium*, vol. 2, pp. II–221–II–224, July 2008.
24. J. Wang, X. Liang, L. Chen, and K. Li, "A Novel Space–Time Coding Scheme Used for MIMO-SAR Systems," *IEEE Geoscience and Remote Sensing Letters*, vol. 12, pp. 1556–1560, 2015.
25. F. He, Z. Dong, and D. Liang, "A Novel Space–Time Coding Alamouti Waveform Scheme for MIMO-SAR Implementation," *IEEE Geoscience and Remote Sensing Letters*, vol. 12, pp. 229–233, 2015.
26. S. Liu, Z. Zhang, and W. Yu, "A Space-Time Coding Scheme With Time and Frequency Comb-Like Chirp Waveforms for MIMO-SAR," *IEEE Journal of Selected Topics in Signal Processing*, vol. 11, pp. 391–403, 2017.
27. J. Kim, M. Younis, A. Moreira, and W. Wiesbeck, "A Novel OFDM Chirp Waveform Scheme for Use of Multiple Transmitters in SAR," *IEEE Geoscience and Remote Sensing Letters*, vol. 10, pp. 568–572, 2013.
28. J. Wang, X. Liang, and L. Chen, "MIMO SAR system using digital implemented OFDM waveforms," in *2012 IEEE International Geoscience and Remote Sensing Symposium*, pp. 7428–7431, July 2012.
29. T. Rommel, R. Rincon, M. Younis, G. Krieger, and A. Moreira, "Implementation of a MIMO SAR Imaging Mode for NASA's Next Generation Airborne L-Band SAR," in *EUSAR 2018; 12th European Conference on Synthetic Aperture Radar*, pp. 1–5, June 2018.
30. J. Wang, L. Chen, X. Liang, C. Ding, and K. Li, "Implementation of the OFDM Chirp Waveform on MIMO SAR Systems," *IEEE Transactions on Geoscience and Remote Sensing*, vol. 53, pp. 5218–5228, 2015.
31. J. Kim, M. Younis, A. Moreira, and W. Wiesbeck, "Spaceborne MIMO Synthetic Aperture Radar for Multimodal Operation," *IEEE Transactions on Geoscience and Remote Sensing*, vol. 53, pp. 2453–2466, 2015.

32. G. Jin, Y. Deng, R. Wang, W. Wang, P. Wang, Y. Long, Z. M. Zhang, and Y. Zhang, "An Advanced Nonlinear Frequency Modulation Waveform for Radar Imaging With Low Sidelobe," *IEEE Transactions on Geoscience and Remote Sensing*, vol. 57, pp. 6155–6168, 2019.
33. G. Jin, K. Liu, Y. Deng, Y. Sha, R. Wang, D. Liu, W. Wang, Y. Long, and Y. Zhang, "Nonlinear Frequency Modulation Signal Generator in LT-1," *IEEE Geoscience and Remote Sensing Letters*, vol. 16, pp. 1570–1574, 2019.
34. H. L. Van Trees, *Optimum array processing*. John Wiley & Sons, 2004.
35. G. Jin, Y. Deng, W. Wang, Y. Zhang, D. Liang, and R. Wang, "A novel spaceborne mimo-sar imaging scheme based on improved ofdm waveforms," *IEEE Geoscience and Remote Sensing Letters*, pp. 1–5, 2020.
36. X. Song, S. Zhou, and P. Willett, "Reducing the waveform cross correlation of mimo radar with space–time coding," *IEEE Transactions on Signal Processing*, vol. 58, pp. 4213–4224, 2010.
37. S. D. Blunt, J. Jakabosky, M. Cook, J. Stiles, S. Seguin, and E. L. Mokole, "Polyphase-coded FM (PCFM) radar waveforms, part II: optimization," *IEEE Transactions on Aerospace and Electronic Systems*, vol. 50, pp. 2230–2241, 2014.

The manufacturer's authorised representative in the EU is Springer Nature Customer Service Centre GmbH, Europaplatz 3, 69115 Heidelberg, Germany. If you have any concerns regarding our products, please contact ProductSafety@springernature.com

Printed and bound by CPI Group (UK) Ltd, Croydon, CR0 4YY
20/10/2026
02236478-0004